Engineering drawing for technicians volume 2

O. Ostrowsky

B.A., M.S.E., P.Eng., Dip. Eng. Design, M.I.E.D., M.I.Plant E., Cert. Ed.
Lecturer responsible for engineering drawing and design,
People's College for Engineering and Science,
Nottingham

Edward Arnold

Preface

This book completely covers the objectives of the Technician Education Council standard units Engineering drawing II (U76/354 and U80/710), with the following additional topics:

a) spur, bevel, and worm-and-wheel gears;
b) general proportions of bolts, studs, nuts, and screws necessary for drawing assemblies;
c) threaded fasteners and keys necessary for the preparation of assembly drawings;
d) consideration of plastics as engineering materials;
e) more detailed work on plain and rolling bearings, their materials, and bearing assemblies for future design work.

More than 160 test questions are included, with 24 self-assessment questions for quick checking of students' progress. All these questions are suitable for phase tests, final tests, classwork, or homework.

Solutions are given to many selected questions, and these include about 100 individual drawing and numerical solutions. In addition, about 15 worked examples are incorporated to make work easier for both the student and the lecturer.

In order to reduce the time-consuming transfer of measurements when preparing certain solutions, and also for testing purposes, 45 g/m^2 paper can be used as an economical substitute for tracing paper.

It was intended to write this book in simple and clear language to ensure that each student could do the additional work on his own at home, if required, without the lecturer's help.

Some material included in *Engineering drawing for technicians volume 1* has been reproduced in this book as it now seems more appropriate as level-2 work.

Acknowledgements

I wish to express thanks to Syd Bullock and other colleagues who read the manuscript and offered helpful criticism. I would also like to thank my wife Catherine, her sister Eileen, and my daughters Sharon and Lisa for their assistance during the preparation of this book.

The following are reproduced from or based on British Standards by kind permission of the British Standards Institution, 2 Park Street, London W1A 2BS, from whom copies of the complete standards may be obtained: fig. 2.3 (BS 3643:part 1:1963), fig. 2.4 (BS 84:1956), fig. 2.6 (BS 5346:1976), fig. 2.7 (BS 1657:1950), fig. 7.9 (BS 4500A:1970), fig. 9.1 (BS 4190:1967 and PD 7300:1979), fig. 9.2 and Table 9.1 (BS 4439:1969), figs 9.6 and 9.7 (BS 4235:part 1: 1972). The general principles of tolerancing discussed on pages 42—4 are based on BS 4500: part 1:1969 and the principles and examples of geometrical tolerancing on pages 46—53 are based on BS 308:part 3:1972.

O. Ostrowsky

Contents

© O. Ostrowsky 1981

First published 1981
by Edward Arnold (Publishers) Ltd
41 Bedford Square, London WC1B 3DQ

ISBN 0 7131 3429 1

British Library Cataloguing in Publication Data
Ostrowsky, O.
     Engineering drawing for technicians.
     Vol. 2.
     1. Mechanical drawing
     I. Title
     604'.2'4   T353

     ISBN 0-7131-3429-1

Text set in 10/12 pt IBM Press Roman by Tek-Art Ltd, and printed in Hong Kong by Wing King Tong Co.Ltd.

1 Loci

The locus of a point is the path traced by the point when it moves in accordance with specified conditions. The plural of locus is *loci*.

1.1 Common loci

Circle

If a point P moves in one plane so that its distance from a fixed point O is constant, then its locus is a circle.

To draw a circle, compasses are set to the required constant distance. With the point of the compass at O (fig. 1.1), the compass lead then traces out the required circle through P_1, P_2, P_3, etc., where $OP_1 = OP_2 = OP_3 = R$, the radius of the circle.

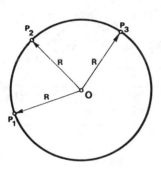
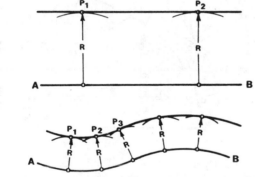

Fig. 1.1 Circle   **Fig. 1.2** Parallel line

Parallel line

If a point P moves in one plane so that its perpendicular distance from a fixed line AB is constant, then its locus is a line parallel to AB.

To draw a parallel line (fig. 1.2):

1. With centres on AB, strike a number of arcs of radius R equal to the required distance between AB and the parallel line.
2. Draw a common tangent to all these arcs. This is the required parallel line.

Perpendicular line

If a point P moves in one plane so that it is equidistant from two fixed points A and B, then its locus is a straight line perpendicular to AB.

To draw a perpendicular line (fig. 1.3):

1. With centres at A and B, strike two arcs each of an arbitrary radius R_1 to intersect at P_1 on either side of AB.
2. With the same centres, strike further pairs of arcs with radii $R_2, R_3, \ldots$, etc. and intersection points $P_2, P_3, \ldots$ on either side of AB. A straight line drawn through $P_1, P_2, P_3, \ldots$, etc. is the required line perpendicular to AB (and is also the bisector of AB).

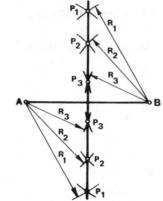
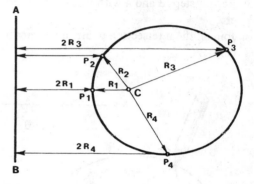

Fig. 1.3 Perpendicular line   **Fig. 1.4** Ellipse

Ellipse – method 1

If a point P moves in one plane so that its distance from a fixed point C and its perpendicular distance from a fixed line AB are always in the same ratio 1:*n*, where *n* is any number greater than 1, then the locus of the point is an ellipse.

To draw an ellipse (fig. 1.4):

1. Taking a distance ratio of, say, 1:2, draw a line parallel to AB and at an arbitrary distance $2R_1$ from it.
2. From centre C, strike an arc of radius R_1 to intersect this line at P_1.
3. Repeat steps 1 and 2 with radii $R_2, R_3, \ldots$, etc. to give intersection points $P_2, P_3, \ldots$, etc.
4. Join all the intersection points by a smooth curve. This curve is the required ellipse.

Ellipse – method 2

If a point P moves in one plane so that the sum of its distances from two fixed points A and B is constant, then its locus is again an ellipse.

If a piece of string of total length equal to AP + PB is fixed with its ends at A and B and is kept taut by a pencil held against it inside the loop so formed, moving the pencil will produce a locus which is an ellipse.

To draw an ellipse (fig. 1.5):

1. For measuring purposes, draw a construction line of total length equal to AP + PB (fig. 1.5(a)).
2. Draw the two fixed points A and B (fig. 1.5(b)).
3. From centre A, strike an arc with radius AP_1.
4. From centre B, strike an arc with radius BP_1, measured from the construction line. This arc intersects the first one at P_1.
5. Repeat steps 3 and 4 with various distances and intersection points P_2, P_3, . . . , etc.
6. Join all the intersection points by a smooth curve. The resulting locus is an ellipse.

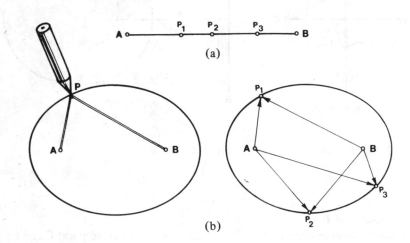

(a)

(b)

Fig. 1.5 Ellipse

Cycloid

A cycloid is the locus of a point on the circumference of a circle which rolls without slip along a fixed straight line.

When a disc with a pencil positioned on its periphery is rolled along a straight edge without slip, the moving pencil will trace the cycloid curve.

The profile of certain gear teeth is based on the cycloid curve.

To draw a cycloid:

1. Draw a circle of radius R, say, with centre O and mark off, say, twelve points equally spaced around its circumference as shown in fig. 1.6.
   Number the points in a clockwise direction, the bottom point being numbered twice (0 and 12).
2. Through each point draw a horizontal construction line.
3. On the construction line passing through O, mark out a distance equal to the circumference of the circle $(2\pi R)$ and divide it into twelve equal parts with points O, O_1, . . . , O_{12}.
4. From centre O, strike an arc of radius R to intersect the construction line from O on the circumference. Call the intersection point P_0.
5. From centre O_1, strike another arc of radius R to give point P_1 where it intersects the construction line from 1 on the circumference.
6. From centres O_2, O_3, . . . , O_{12}, strike arcs in the same way to give points P_2, P_3, . . . , P_{12}.
7. Join points P_0 to P_{12} by a smooth curve. The resulting locus is the cycloid.

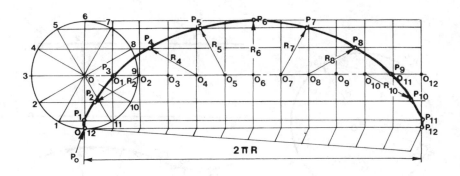

Fig. 1.6 Cycloid

Epicycloid and hypocycloid

An *epicycloid* is the locus of a point on the circumference of a circle which rolls without slip along the *outside* of a fixed base circle.

A *hypocycloid* is the locus of a point on the circumference of a circle which rolls without slip along the *inside* of a fixed base circle.

The profile of certain gear teeth is based on a combination of epicycloid and hypocycloid curves, as shown in fig. 1.7(b).

To draw an epicycloid or hypocycloid (fig. 1.7(a)):

1. Strike an arc of the appropriate radius for the fixed base circle and draw the rolling circle of radius R touching it – above the arc for the epicycloid, below it for the hypocycloid.

2

2. Mark off, say, twelve equally spaced points around the circumference of the rolling circle and number them as shown — clockwise for the epicycloid, anticlockwise for the hypocycloid.
3. With centres at the centre of the fixed base circle, draw construction arcs through these points and through the centre of the circle, O.
4. Along the arc of the fixed base circle, mark off a length equal to the circumference of the rolling circle ($2\pi R$). This length subtends an angle θ at the centre of the base-circle arc.
5. Divide this angle θ into twelve equal parts by lines passing through the construction arcs. On the construction arc passing through the centre of the rolling circle, O, number the intersection points $O_1, O_2, \ldots, O_{12}$ as shown.
6. From centre O_1, strike an arc of radius R (the rolling-circle radius) to give point P_1 where it intersects the construction arc passing through point 1 on the rolling circle.
7. From centre O_2, strike another arc of radius R to give point P_2 where it intersects the construction arc passing through point 2 on the rolling circle.
8. From points $O, O_3, O_4, \ldots, O_{12}$, strike arcs in the same way to give points $P_0, P_3, P_4, \ldots, P_{12}$.
9. Join points P_0 to P_{12} by a smooth curve. The resulting locus will be the epicycloid or hypocycloid as appropriate.

Involute

An involute is the locus of a point on a straight line which rolls without slip around a circle.

If a length of string with a pencil attached to its free end is unwound under tension from around the circumference of a fixed disc, then the moving pencil will trace an involute curve.

Part of the involute curve is used for gear-tooth profiles, as shown in fig. 1.8(b); it has many advantages over the cycloid curves.

To draw an involute:
1. Draw the base circle and mark off, say, twelve equally spaced points around its circumference, numbering the points as shown in fig. 1.8(a).
2. At point 1, draw a tangent of length equal to one twelfth of the circumference of the circle. The 'free' end of the tangent is point P_1.
3. At point 2, draw a tangent of length equal to two twelfths of the circumference of the circle, to give point P_2.
4. In the same way, from points $3, 4, \ldots, 12$ draw tangents of length $3/12$, $4/12, \ldots, 12/12$ of the circumference of the circle respectively, to give points $P_3, P_4, \ldots, P_{12}$.
5. With the aid of a French curve, join points P_1 to P_{12} by a smooth curve. The resulting locus will be the involute.

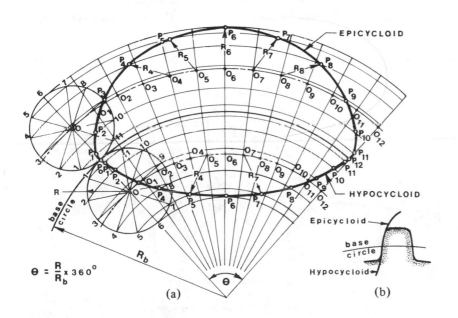

$$\Theta = \frac{R}{R_b} \times 360°$$

(a)

(b)

Fig. 1.7 Epicycloid and hypocycloid

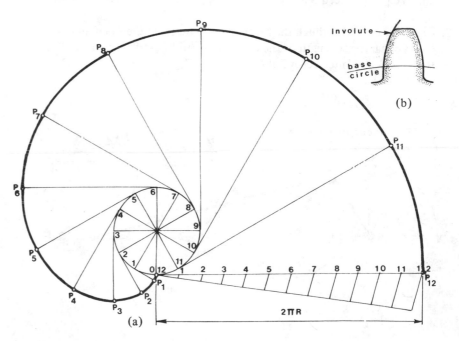

(b)

(a)

Fig. 1.8 Involute

3

1.2 Mechanisms

A mechanism is a machine or part of a machine consisting of a system of moving parts. Any rigid piece of the mechanism pivoted at the ends is called a *link*. Arrangements of links, pivots, and slides are often used to convert circular motion into reciprocating or oscillating motion, and vice versa.

When designing a mechanism to perform a specific task, the loci of various points on the mechanism must be drawn to determine the velocities, accelerations, and forces involved in the motion produced. It is also necessary to study the movements of the links so that clearances may be checked and safety guards or shields may be designed to protect the users of the machine.

Two very common types of mechanism are the slider-crank and the four-bar chain.

Slider-crank

The slider-crank is a simple mechanism in which rotary motion is converted to linear motion, or vice versa. Figure 1.9(a) shows a crank AB which rotates about A and is joined by the connecting rod BC to the piston (slider) C which slides along the axis AC.

To plot the locus of a point P on the mechanism (fig. 1.9(b)):

1. Draw a circle of radius equal to the crank length AB and mark off, say, twelve points equally spaced on its circumference as shown. Number the points B_1, B_2, ..., B_{12}.
2. Draw the axis AC through the centre of the circle and the slider and, with radius equal to the length of the connecting rod BC, strike arcs from B_1, B_2, ..., B_{12} to intersect AC at C_1, C_2, ..., C_{12}.

3. Join $B_1 C_1$, $B_2 C_2$, ..., $B_{12} C_{12}$ to give the positions of the connecting rod BC for twelve different positions of the crank AB.
4. With the same centres B_1, B_2, ..., B_{12} and radius BP, strike arcs to intersect $B_1 C_1$, $B_2 C_2$, ..., $B_{12} C_{12}$ at P_1, P_2, ..., P_{12}.
5. Join points P_1, P_2, ..., P_{12} by a smooth curve. This is the locus of point P.

Alternatively, a trammel method may be used. The trammel, which may be a piece of paper with a straight edge on which the relevant points are marked, represents the connecting rod and its movements are plotted. The trammel method enables a large number of points to be obtained quickly, and so is widely used by designers in industry.

To plot the locus of a point P on the mechanism using a trammel:

1. On an axis AC with centre A, draw a circle of radius equal to the crank length AB.
2. Mark points BPC on the edge of the paper to represent the connecting rod.
3. Place the paper in various positions such that point B always touches the circumference of the circle and point C is always on the axis AC. For each position, mark the position of P.
4. Join the successive positions of P to obtain the required locus.

Four-bar chain

The four-bar chain is a simple mechanism which consists of two cranks, AB and CD, joined by a rod BC. The fourth link is between the two fixed pivots A and D, as shown in fig. 1.10(a).

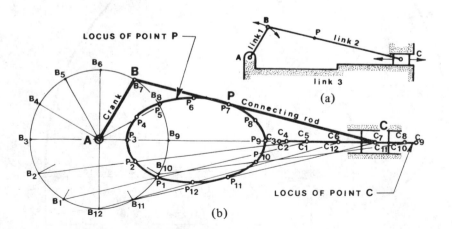

Fig. 1.9 Slider-crank mechanism

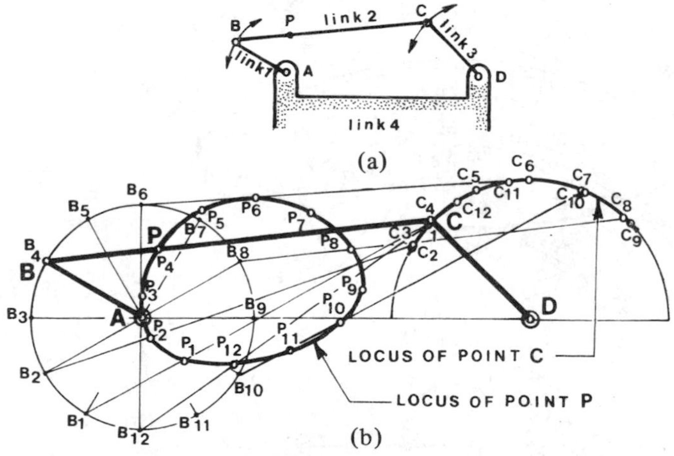

Fig. 1.10 Four-bar-chain mechanism

To plot the locus of a point P (fig. 1.10(b)):

1. Draw a circle of radius AB with centre A and another of radius DC with centre D.
2. Mark off, say, twelve equally spaced points on the circle of radius AB to correspond to twelve different positions of the crank AB. Number the points $B_1, B_2, \ldots, B_{12}$ as shown.
3. With $B_1, B_2, \ldots, B_{12}$ as centres, strike arcs of radius equal to the length of the connecting rod BC to intersect the circumference of the circle of radius DC at points $C_1, C_2, \ldots, C_{12}$.
4. Join $B_1 C_1, B_2 C_2, \ldots, B_{12} C_{12}$ to give the positions of the connecting rod BC for twelve different positions of the crank AB.
5. Again with centres $B_1, B_2, \ldots, B_{12}$, strike arcs of radius BP to intersect $B_1 C_1, B_2 C_2, \ldots, B_{12} C_{12}$ at $P_1, P_2, \ldots, P_{12}$.
6. Join points P_1 to P_{12} by a smooth curve. This is the locus of point P. Alternatively, a trammel method may be used.

Sliding link

If a ladder AB is propped against a wall and the bottom end slides outwards from B_1 to B_2, then the top of the ladder will slip correspondingly from A_1 to A_2, since the length of the ladder remains constant.

A sliding link is one which is free to move in one plane such that one end of the link is always in contact with a line OY and the other end is always in contact with another line OX.

To plot the locus of a point P on the link (fig. 1.11):

1. With centres $A_1, A_2, \ldots$, etc. on line OY and radius equal to the length of the link AB, strike a number of arcs to intersect OX at $B_1, B_2, \ldots$, etc.
2. Join $A_1 B_1, A_2 B_2, \ldots$, etc. to obtain various positions of the link.
3. Again with centres $A_1, A_2, \ldots$, etc., strike arcs of radius AP to intersect $A_1 B_1, A_2 B_2, \ldots$, etc., at $P_1, P_2, \ldots$, etc.
4. Join the points $P_1, P_2, \ldots$, etc. with a smooth curve. This is the required. locus.

Quick-return mechanism

The quick-return mechanism shown in fig. 1.12 is used to reduce the time wasted during the non-cutting return stroke of a shaping machine.

A pinion drives a gear wheel with a uniform angular velocity. Attached to the gear wheel is a pin which rotates with it and at the same time moves a sliding block up and down inside the slotted link, which in turn oscillates about its pivot.

The shaping-machine ram, which is joined to the top of the slotted link by means of a connecting link, makes its cutting stroke while the pin is travelling through the larger arc, of about 240°. The non-cutting return stroke is made while the pin travels through the smaller arc, of about 120°. Thus the ram moves back twice as fast as it moves forward on its cutting stroke.

The length of the ram stroke can be adjusted by changing the radial distance of the pin from the centre of the gear wheel.

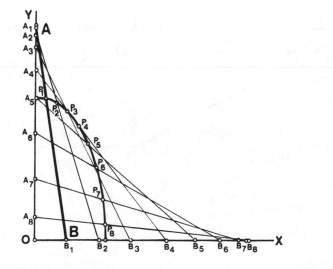

Fig. 1.11 Sliding link

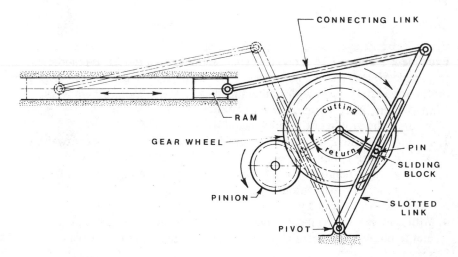

Fig. 1.12 Quick-return mechanism

5

1.3 Worked examples

Example 1 If a point P moves in one plane so that its distances from two fixed points A and B are always in the same ratio 2:1, then its locus is a circle.

To draw a circle where A and B are 60 mm apart (fig. 1.13):

1. Strike an arc with radius R_1 of 20 mm from point B and another arc with radius $2R_1$ from point A to intersect the first arc at P_1.
2. Repeat the procedure with different radii, still in the ratio 2:1, to give intersection points $P_2, P_3, \ldots$, etc.
3. Join all the intersection points with a smooth curve. The resulting locus is the circumference of a circle as shown (to a reduced scale).

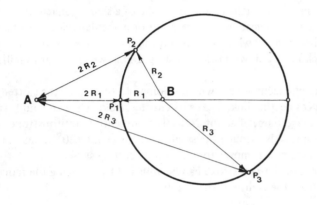

Fig. 1.13 Circle

Example 2 Figure 1.14 shows a mechanism in which a rod ED is pin-jointed to a crank AB at B and is constrained to pass through a fixed point C. Design a profile for a safety guard with a minimum clearance of 12 mm, where AC = 100 mm, AB = 40 mm, ED = 180 mm, and EB = 20 mm.

To draw the profile of the safety guard:

1. After positioning A and C, draw a circle of radius AB and mark off twelve equally spaced points.
2. Through these points, draw construction lines representing the rod ED, making sure they pass through C.
3. Draw the loci of points E and D.
4. With centres on these locus curves and clearance radius of 12 mm, strike arcs to obtain the tangential profile of a safety guard as shown (to a reduced scale).

Alternatively, a trammel method may be used.

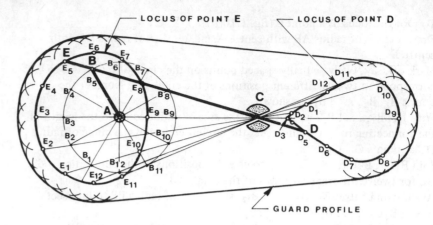

Fig. 1.14 Mechanism

Example 3 Figure 1.15 shows a wrapping-machine linkage pivoted at D and A and pin-jointed at C, B, and E. Link EP, length 100 mm, is constrained to pass through the point F, and link AB, length 50 mm, oscillates about A between B_1 and B_2. Design an outline of a safety guard with a minimum clearance of 10 mm.

To draw the outline of a safety guard:

1. Draw the locus of point E (fig. 1.10). Using this locus, draw the required locus of point P.
2. Draw the outline of a safety guard as shown (to a reduced scale).

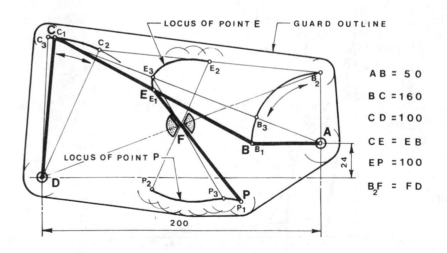

AB	= 50
BC	= 160
CD	= 100
CE	= EB
EP	= 100
B_2F	= FD

Fig. 1.15 Wrapping-machine linkage

6

1.4 Test questions

1. Define the locus of a moving point.
2. Relate the locus definition to the circle, ellipse, and cycloid.
3. Define the epicycloid, hypocycloid, and involute and explain for what purposes these curves are used in engineering.
4. Explain the importance for engineering designers of determining the locus of a point on a moving mechanism.
5. Construct a perpendicular line to bisect a line AB 47 mm long.
6. Draw a line AB and a point C 60 mm from it. Then draw the locus of point P which moves in such a way that it is always at one chosen distance from the point C and at double the chosen distance measured perpendicularly from the line AB.
7. Draw the locus of a point P which moves so that the sum of its distances AP + PB from two fixed points A and B remains constant at 120 mm. The distance between A and B is 75 mm.
8. Draw a cycloid of a rolling circle 80 mm diameter.
9. Draw the epicycloid and hypocycloid for a rolling-circle radius of 30 mm and a base-circle radius of 90 mm.
10. a) Define the involute and explain why this curve is important in engineering design.
    b) Draw the involute on a base circle of 50 mm diameter.
11. Draw the locus of point P for the slider-crank mechanism shown in fig. 1.16.
12. Draw the locus of point P for the four-bar chain shown in fig. 1.17.
13. Figure 1.18 shows a pin-jointed mechanism. The cranks AB and CD revolve about A and C at the same speed.
    Draw the locus of the points E and F and then the outline of a safety guard with a minimum clearance of 10 mm.
14. Draw the machine mechanism shown in fig. 1.19 and plot the locus of point E for a complete revolution of the crank AB.
    Draw also the guard outline with a minimum clearance of 12 mm.
15. A mechanism used in a textile machine is shown in fig. 1.20. Crank AC rotates about A. Connecting rod BE is pin-jointed at B, with E constrained to reciprocate vertically. DF is pin-jointed at D but is free to slide through the swivelling guide C at all times.
    Draw the loci of points D and F and construct the profile of a suitable guard to enclose the mechanism with a minimum clearance of 15 mm.
16. Draw the locus of point E of the mechanism shown in fig. 1.21, where BE is constrained to pass through the tee-piece at C.
17. Figure 1.22 shows a mechanism pivoted at A and D and pin-jointed at C, E, and B. The link EG is free to slide through the swivelling guide F at all times. Plot the path of points E and G and construct a safety guard with a minimum clearance of 16 mm.

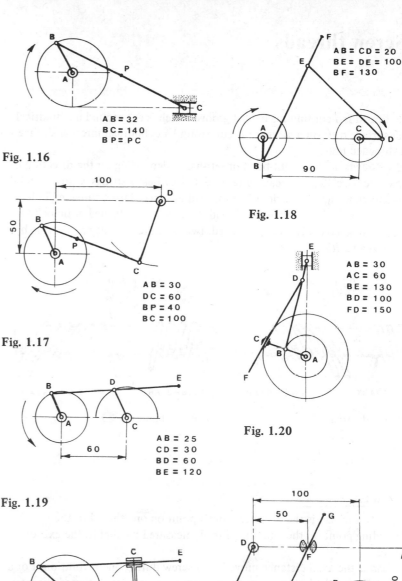

AB = 32
BC = 140
BP = PC

Fig. 1.16

AB = CD = 20
BE = DE = 100
BF = 130

Fig. 1.18

100

50

AB = 30
DC = 60
BP = 40
BC = 100

Fig. 1.17

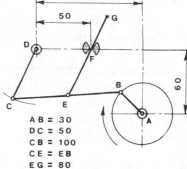

AB = 30
AC = 60
BE = 130
BD = 100
FD = 150

Fig. 1.20

AB = 25
CD = 30
BD = 60
BE = 120

Fig. 1.19

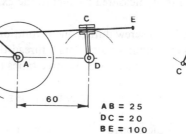

AB = 25
DC = 20
BE = 100

Fig. 1.21

100
50

AB = 30
DC = 50
CB = 100
CE = EB
EG = 80

60

Fig. 1.22

7

2 Screw threads

A screw thread is a continuous helical groove which is cut around a cylindrical external surface to form a screw or is cut around a cylindrical internal surface to form a threaded hole.

Screw threads may be right-hand or left-hand, depending on the direction of the helix (see page 12). This can be represented by heavy strings wound around a rod as shown in fig. 2.1. A right-hand thread advances into a threaded hole when turned clockwise; a left-hand thread advances when turned anticlockwise.

When a quick axial advance is required, two or more threads are cut side by side to form a *multi-start* thread.

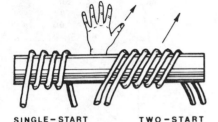 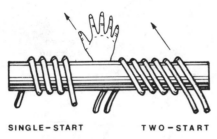

(a) Left-hand thread   (b) Right-hand thread

Fig. 2.1

2.1 Thread terms (fig. 2.2)

The *pitch* of a thread is the distance from a point on one thread to the corresponding point on the adjacent thread, measured parallel to the axis of the thread.

The *lead* is the axial distance moved by a screw during one revolution. For a single-start thread the lead is equal to the pitch; for a two-start thread the lead is equal to twice the pitch; for a three-start thread the lead is three times the pitch; and so on (see page 14).

The *crest* of a thread is the most prominent part of an external or internal thread.

The *root* of a thread is the bottom of the groove of an external or internal thread.

The *flank* of a thread is the straight side connecting the crest to the root.

The *thread angle* is the angle between the flanks of the thread.

The *major diameter,* or outside diameter, is the greatest diameter of a thread.

The *minor diameter,* or root diameter, is the smallest diameter of a thread.

The *effective diameter,* or pitch diameter, is the diameter between the two opposite pitch lines, a pitch line being a line which intersects the flanks of the threads parallel to the axis such that the widths of the threads and of the spaces between the threads are equal.

The *thread depth* is the radial distance between the crests and the roots.

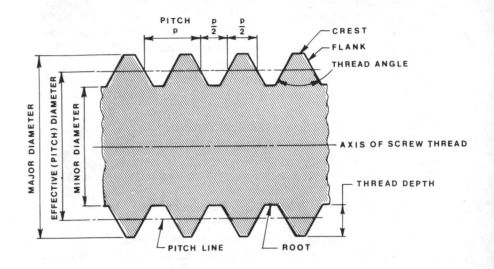

Fig. 2.2 Thread terms

2.2 Thread forms

There are two main classes of thread forms: *vee threads,* which are principally used for fastening and adjusting purposes, and *square threads,* which are used for transmitting forces (power transmission).

2.3 The ISO metric thread (fig. 2.3)

The ISO (International Organisation for Standardisation) metric thread is a vee-form thread. There are two types of ISO metric thread:

a) *fine-pitch-series threads,* which are mainly used for special applications such as for thin-walled components, fine adjustments on machine tools, etc.;

b) *coarse-pitch-series threads,* which are suitable for metric fasteners and for general-purpose applications.

8

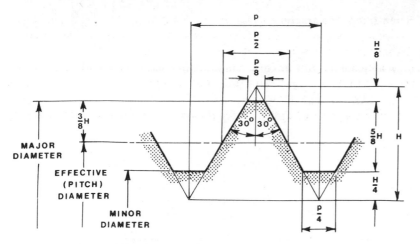

Fig. 2.3 Basic form of ISO metric thread

Classes of fit

There are three classes of fit between external (bolt) threads and internal (nut) threads for different engineering uses: close, medium, and free.

The close fit　This fine-tolerance fit is applied to threads used for very high-quality precision work and requires very thorough inspection.

The medium fit　This is suitable for most general engineering purposes.

The free fit　This coarse-tolerance fit is applied where a quick and easy assembly is needed, with threads occasionally becoming dirty and slightly damaged.

Designation of ISO metric threads

The complete designation of ISO metric screw threads is shown in the following examples:

a) For an internal thread:

　　M20 x 2.5 − 6H

　　　　　　└── Thread tolerance-class symbol
　　　　└── Pitch in millimetres
　　└── Nominal size (major diameter) in millimetres
　└── Symbol for ISO metric thread

b) For an external thread:

　　M20 x 2.5 − 6g

c) For a pair of threaded parts:

　　M20 x 2.5 − 6H/6g

On a drawing, the thread tolerances are usually not indicated and the ISO metric screw threads are designated as, for instance, M20 x 1.5 for the fine-pitch series and M20 x 2.5 for the coarse-pitch series, or simply as M20 when the coarse threads are specified.

2.4 Other thread forms

British Standard Pipe (BSP) thread (fig. 2.4)

This Whitworth standard-form thread has a relatively fine pitch so that both internal and external threads may be cut on thin tubing and piping. When used for specific high-pressure applications, the BSP threads are cut in a tapered rather than a cylindrical surface.

BSP threads are designated by the size of the internal diameter (bore) of the pipe, instead of by the thread major diameter as is the case for all other types of thread.

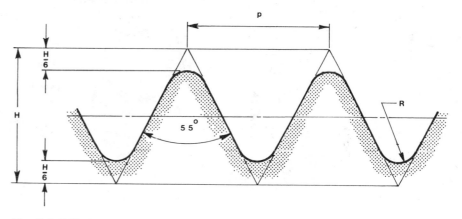

Fig. 2.4 BSP thread

Square thread (fig. 2.5)

The square thread form is used mainly to transmit forces in both axial directions. Since its flanks are normal to its axis, it offers less frictional resistance to motion than does a vee thread. The main applications of this thread include valve spindles, machine leadscrews, vice screws, and screw-jacks.

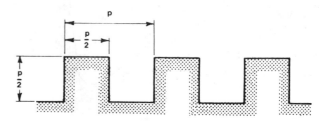

Fig. 2.5 Square thread

9

ISO metric trapezoidal screw thread (fig. 2.6)

This thread is a modified form of square thread. It is stronger, because of the wider base, and easier to cut, due to its taper. The trapezoidal thread is widely used to transmit forces in valve spindles, lathe leadscrews, etc., but its inclined flanks give rise to frictional resistance. When a split clamping nut is used, as on the leadscrew of a lathe, both halves of the nut will engage easily on the tapered sides of this thread.

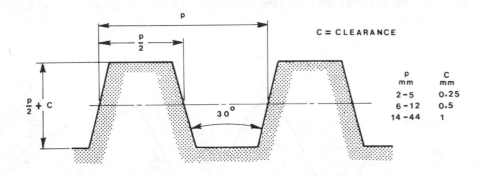

Fig. 2.6 ISO metric trapezoidal screw thread

Buttress thread (fig. 2.7)

This strong thread form offers less frictional resistance than the vee-type threads and is designed to withstand heavy forces applied in one direction. Its main applications include quick-release vices, slides, and presses.

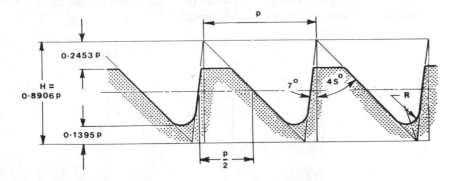

Fig. 2.7 Buttress thread

2.5 Screw-thread constructions

Square thread

To draw a square thread form of 6 mm pitch to a scale of 10:1 (fig. 2.8):

1. Draw two horizontal lines half a pitch (30 mm) apart and then a number of vertical lines half a pitch (30 mm) apart.
2. Complete the required thread profile by drawing the thick outlines.

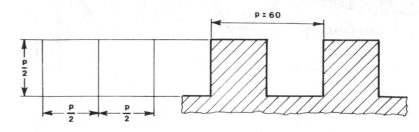

Fig. 2.8 Square thread

ISO metric trapezoidal thread

To draw an ISO metric trapezoidal thread form of 6 mm pitch to a scale of 10:1 (fig. 2.9):

1. From a horizontal line AB representing the pitch line, draw say six flanks half a pitch (30 mm) apart and inclined at 15° to the vertical to give the thread angle of 30°.
2. On each side of the pitch line, draw a parallel line a quarter of a pitch (15 mm) plus half a clearance (2.5 mm) away.
3. Complete the required thread profile by drawing the thick outlines.

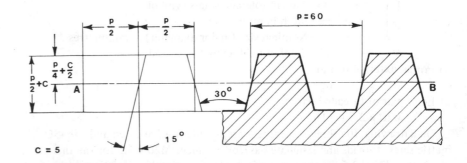

Fig. 2.9 ISO metric trapezoidal thread

Buttress thread

To draw a buttress thread of 6 mm pitch to a scale of 10:1 (fig. 2.10):

1. From a horizontal line AB, draw say four vertical lines a pitch (60 mm) apart and then draw the flanks inclined at 7° and 45° to the vertical.
2. Join the apex points to obtain the height H of 0.8906 x 60 mm = 53.44 mm and then draw two horizontal lines, one at 0.2453 x 60 mm = 14.72 mm from AB and the other at 0.1395 x 60 mm = 8.37 mm from CD.
3. Bisect the angles marked E and F to obtain the intersection point G, and transfer this point horizontally to the corresponding positions for adjacent teeth.
4. With centres G, draw tangential arcs and complete the required thread profile by drawing the thick outlines.

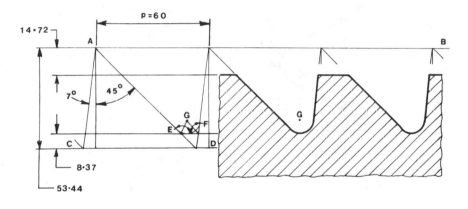

Fig. 2.10 Buttress thread

ISO metric thread

a) To draw an *external* (bolt) thread M64 x 6 to a scale of 10:1 (fig. 2.11);

1. From a horizontal line AB, draw say seven vertical lines half a pitch (30 mm) apart and then draw the flanks inclined at 30°.
2. Join the apex points to obtain the height H and divide this into first six and then eight equal parts. (see volume 1, page 46). At $H/8$ from the top and $H/6$ from the bottom, draw horizontal lines.
3. At $H/6$, bisect the angles marked C and D to obtain the intersection point E, which should lie on the vertical construction line as shown. Transfer the point E horizontally to the corresponding positions for adjacent teeth.
4. With centres E, draw tangential root arcs and complete the required thread profiles by drawing the thick outlines.

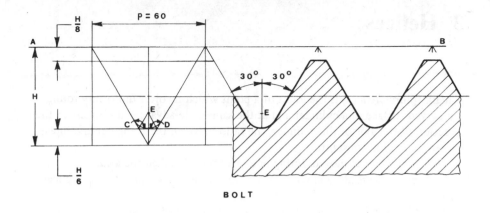

Fig. 2.11 ISO metric external thread

b) To draw an *internal* (nut) thread M64 x 6 to a scale of 10:1 (fig. 2.12):

1. From a horizontal line AB, draw say seven vertical lines half a pitch (30 mm) apart and then draw the flanks inclined at 30° to give the thread angle of 60°.
2. Join the apex points to obtain the height H and divide this into eight equal parts (see volume 1, page 46). At $H/8$ from the top and $H/4$ from the bottom, draw horizontal lines.
3. Complete the required thread profile by drawing the thick outlines.

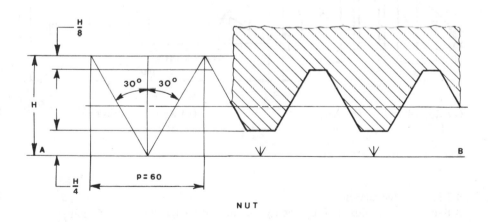

Fig. 2.12 ISO metric internal thread

11

3 Helices

A helix (plural *helices*) is the locus of a point which revolves uniformly round the curved surface of a cylinder and at the same time advances uniformly in the direction of the axis of the cylinder, the ratio between these angular and linear movements being constant.

The axial distance moved during one revolution is called the *lead*.

Coil springs, threads, and worm gears are common examples of the application of helices.

The helix is right-handed when the cylinder is viewed axially and the point moves in a clockwise direction and away from the observer, fig. 3.1(a).

The helix is left-handed when the point moves in an anticlockwise direction and away from the observer, fig. 3.1(b).

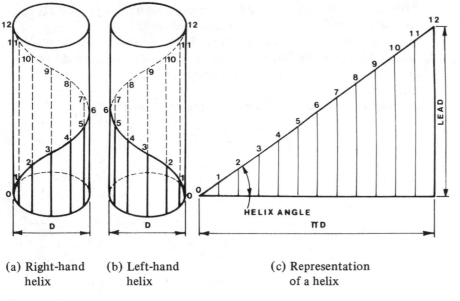

(a) Right-hand helix   (b) Left-hand helix   (c) Representation of a helix

Fig. 3.1 Helix

3.1 Helix construction

A helix can be represented by the hypotenuse of a triangular piece of paper (fig. 3.1(c)) whose height corresponds to the lead of the helix and whose base is wound around the circumference of the cylinder as shown in figs 3.1(a) and (b).

To construct a helix (fig. 3.2)
1. Draw the front and plan views of the base cylinder.
2. On the plan view, mark say twelve equally spaced points, numbering them anticlockwise for a right-hand helix and clockwise for a left-hand helix as shown.
3. From these points, project vertical lines from the plan to the front view.
4. Divide the lead into the same number of equal parts as the plan (twelve) and number the points as shown.
5. Project horizontally from these points to intersect the corresponding vertical projection lines at points $P_1, P_2, \ldots, P_{12}$.
6. With the help of a French curve, join all the intersection points $P_1, P_2, \ldots, P_{12}$ with a smooth curve, This is the required helix.

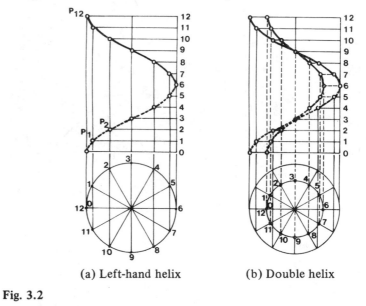

(a) Left-hand helix   (b) Double helix

Fig. 3.2

To draw a screw thread a double helix is constructed, the larger helix representing the crest of the thread and the smaller one the root of the thread as shown in fig. 3.2(b).

To draw a vee thread (fig. 3.3)
1. Draw the front view of the construction cylinder, mark off the lead and divide it into, say, twelve equal parts numbered as shown.
2. Draw the circle for the end view of the construction cylinder, corresponding to the crest of the thread.
3. Draw the thread angle of $60°$ on the front view and project the root of the thread horizontally to the end view.

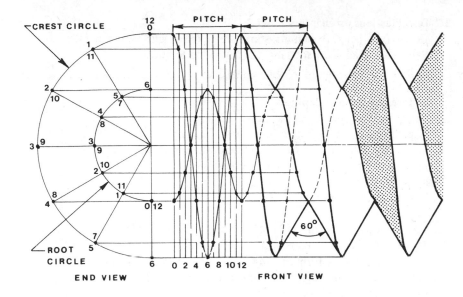

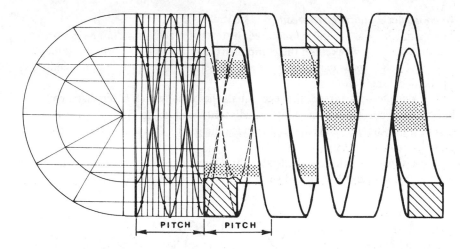

Fig. 3.3 Right-hand vee thread

4. Draw the root circle and divide this and the crest circle into twelve parts as shown. Number the intersection points around the circumference, anticlockwise for a right-hand thread and clockwise for a left-hand thread.

5. From the crest circle, project these points horizontally on to the front view and mark the points where the projector from a particular number on the end view intersects the vertical line with the same number on the front view. These points give the larger helix.

6. Repeat step 5 from the root circle to give the points for the smaller helix, starting from the required point.

7. Complete the front view by drawing smooth curves through the visible parts.

By marking off and dividing several leads along the front view of the construction cylinder and projecting horizontally from the end view through all of them, a greater length of thread may be drawn.

To draw a square thread (fig. 3.4(a))

The same procedure as given above for the vee thread is followed, except that a square of side equal to the half-pitch and the corresponding helices from four corners are drawn. By choosing the correct helix, a right-hand or a left-hand thread may be obtained.

(a) Right-hand square thread   (b) Left-hand square-section spring

Fig. 3.4

To draw a square-section spring (fig. 3.4(b))

Most coiled springs are formed on cylinders, therefore a square-section spring is drawn using the method given above for a square thread but with the cylinder removed, as shown.

To draw a round-section spring (fig. 3.5)

1. Draw the helix corresponding to the centre line of the round spring wire.
2. With centres on the helix, draw a number of circles of the same diameter as the spring wire.
3. Draw the best smooth curves tangential to these construction circles.

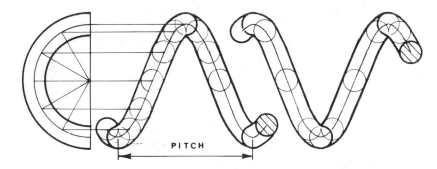

(a) Left-hand spring   (b) Right-hand spring

Fig. 3.5

13

To construct a multi-start thread (fig. 3.6)

When a quick axial advance is required, instead of machining a large-size thread and thus possibly weakening the component, a smaller multi-start thread may be used. Such threads can be single-start, two-start, three-start, etc., depending on the helix angles used.

For a single-start thread, the lead equals the pitch, which is the distance between the corresponding points on adjacent threads (fig. 3.6(a)).

For a two-start thread, the lead is twice the pitch. This means that two threads are cut side by side, fig. 3.6(b).

For a three-start thread, the lead is three times the pitch, with three threads cut side by side, fig. 3.6(c); and so on.

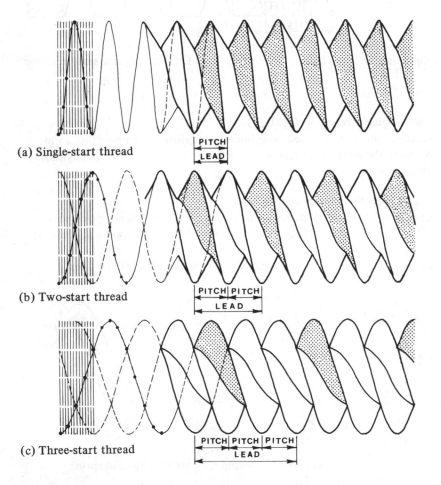

(a) Single-start thread

(b) Two-start thread

(c) Three-start thread

Fig. 3.6

3.2 Test questions on chapters 2 and 3

1. Define (a) a helix, (b) a multi-start thread, (c) a right-hand thread.
2. Sketch a thread form and label the following features:
   a) major, pitch, and minor diameters;
   b) lead and pitch for a single-start thread;
   c) thread angle and depth;
   d) crest, flank, and root of a thread.
3. Explain the meaning of the thread designation M5 x 0.8 − 6H/6g.
4. Sketch the following thread forms, label their important features, and indicate their general use in engineering:
   a) a BSP thread,
   b) a buttress thread,
   c) a square thread.
5. Construct the basic form of a M5 x 0.8 external (bolt) thread to a scale of 100:1 and label the important thread features.
6. Construct the basic form of a M4 x 0.7 internal (nut) thread to a scale of 100:1 and label the important thread features.
7. Construct the basic forms of the following threads to a scale of 50:1, using a pitch of 1 mm:
   a) a BSP thread,
   b) an ISO metric trapezoidal screw thread,
   c) a buttress thread.
8. Explain the difference between a pitch and a lead.
9. Construct three complete turns of a right-hand single-start vee thread to a scale of 10:1 for a pitch of 3.6 mm, a major diameter of 11 mm, and a thread angle of 60°. (See page 13, fig. 3.3.)
10. Construct four complete turns of a right-hand two-start square thread with a pitch of 36 mm and a major diameter of 108 mm.
11. Construct four complete coils of a left-hand square-section spring with a pitch of 36 mm and a pitch diameter of 96 mm. (See page 13, fig. 3.4.)
12. Construct six complete coils of a right-hand round-section spring. Diameter of the spring wire to be 10 mm, pitch (effective) diameter 100 mm, and pitch 48 mm.
13. Construct six complete turns of a right-hand three-start vee thread to a scale of 10:1 for a pitch of 3.6 mm, a major diameter of 12 mm, and a thread angle of 60°.
14. Construct six complete turns of a left-hand three-start square thread with a pitch of 36 mm and a major diameter of 100 mm.
15. Construct five complete turns of a left-hand two-start ISO metric trapezoidal screw thread to a scale of 10:1 for a pitch of 4 mm and a major diameter of 12 mm. Thread depth to be equal to half a pitch, and thread angle to be 30°.

4 Cams

A *cam* is a component which may rotate, oscillate, or reciprocate and is shaped so that it imparts motion to another component, called a *follower,* which may reciprocate in a guide or oscillate about a pivot.

Sometimes a restraining spring is used to ensure that the follower is kept in contact with the cam.

Cams are used in internal-combustion engines to operate valves, in packaging and printing machinery, in machine tools, and in many other industrial applications.

4.1 Main types of cam

The wedge cam (fig. 4.1)
This very simple cam reciprocates horizontally and causes a follower, which is in constant contact with the cam profile, to reciprocate vertically in its guide.

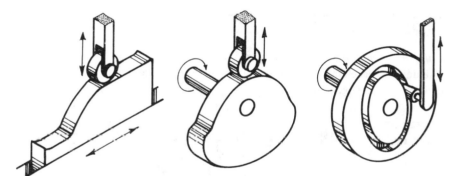

Fig. 4.1 Wedge cam   Fig. 4.2 Disc cam   Fig. 4.3 Face cam

The disc or plate cam (fig. 4.2)
This rotary cam is also known as a *radial* or *edge* cam. It is made of a flat plate with an edge profile to transmit the required motion to a follower.

The face cam (fig. 4.3)
In its flat surface, this rotary cam has a groove machined within which the roller follower is constrained to move. Due to the positive motion of the follower, there is no need for a restraining spring.

The cylindrical cam (fig. 4.4)
In its curved surface, this rotary cam has a groove machined within which the roller follower is constrained to move. The reciprocating positive motion of the follower is parallel to the cam axis.

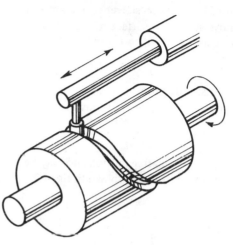 

Fig. 4.4 Cylindrical cam   Fig. 4.5 End cam

The end cam (fig. 4.5)
This cylindrical cam has its end machined to the required shape.

4.2 Main types of cam follower
There are three main types of follower: knife-edge, flat, and roller.

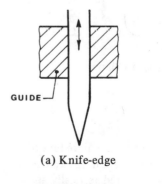  

GUIDE

(a) Knife-edge   (b) Flat   (c) Roller

Fig. 4.6 Cam followers

The knife-edge follower (fig. 4.6(a))

This has the advantage that it can follow any complicated cam profile. However, it is not often used as it wears rapidly, due to high pressures and sliding friction.

The flat follower (fig. 4.6(b))

This cannot be used for concave cam profiles. It wears slower than a knife-edge follower, since the points of contact move across the surface of the follower according to the changing profile of the cam.

The roller follower (fig. 4.6(c))

This has the advantage that wear is minimised, due to rolling rather than sliding friction. The cam profile must not incorporate any concave forms with a radius smaller than the radius of the roller.

The cam follower may have a reciprocating or oscillating motion and may be positioned in line with the cam centre line or be offset, as shown in fig. 4.7.

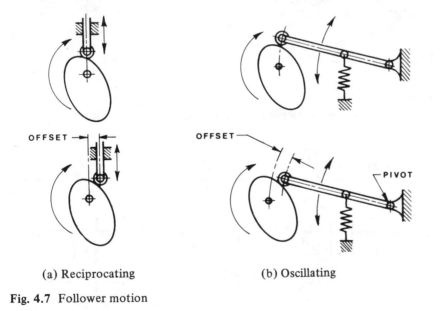

(a) Reciprocating   (b) Oscillating

Fig. 4.7 Follower motion

4.3 Types of follower motion

To design the profile of a rotary cam in order to impart a required motion to a follower, it is very convenient to draw first a displacement diagram. On this diagram, the linear displacement of the follower is plotted vertically against angular displacement for a complete rotation of the cam.

There are three fundamental types of motion which may be imparted to followers: uniform velocity, uniform acceleration and retardation, and simple harmonic motion.

Uniform velocity (fig. 4.8)

The displacement diagram will be a straight sloping line, as the displacement is proportional to the angle turned through, and equal angles turned through by the cam will produce equal increments of rise or fall of the follower.

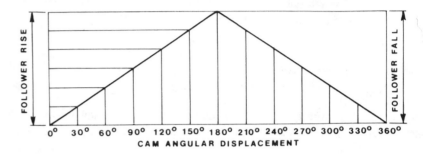

Fig. 4.8 Uniform velocity

Uniform acceleration and retardation

The uniform acceleration and retardation curves are parabolic in form. To construct them (fig. 4.9):

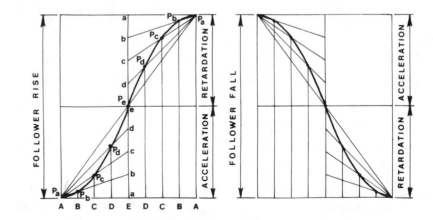

Fig. 4.9 Uniform acceleration and retardation

16

1. Divide the rise of the follower and the angular displacement of the cam into the *same* number of equal parts by points A, B, . . . , etc. and a, b, . . . , etc. as shown.
2. For the acceleration curve, draw construction lines radiating from the *initial* point A to a, b, . . . , etc., and for the retardation curve draw construction lines radiating from the *final* point A to a, b, . . . , etc.
3. Mark point P_b where the line Ab intersects the vertical through B, point P_c where the line Ac intersects the vertical through C, etc. The smooth curves through A, P_b, P_c, . . . , etc. will be the required acceleration and retardation curves.

Simple harmonic motion (s.h.m.)

This curve is a sine curve and it represents a motion similar to that of a swinging pendulum in a clock.

To construct the s.h.m. curve (fig. 4.10):

1. Draw a semicircle of diameter equal to the follower displacement. Divide it into, say, six equal angular parts and project horizontal construction lines from the intersection points thus obtained.
2. Divide the cam displacement into the same number of equal parts (six) and project verticals.
3. Join the corresponding points of intersection of the vertical and horizontal lines to obtain the smooth s.h.m. curve.

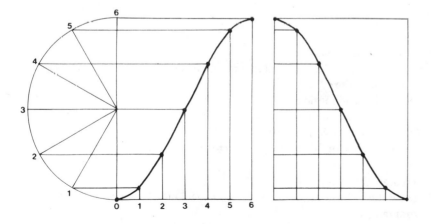

Fig. 4.10 Simple harmonic motion (s.h.m.)

4.4 Construction of cam profiles

1. Figure 4.11(a) shows a follower displacement diagram (F.D.D.) for a wedge cam.
2. Figure 4.11(b) shows that the profile of a wedge cam is identical to the follower displacement diagram when a knife-edge follower is used.
3. Figure 4.11(c) shows that the follower displacement diagram is identical to the locus of the centre of a roller follower. The cam profile is drawn tangential to construction circles representing the roller in various positions.
4. Figure 4.11(d) shows that the follower displacement diagram is identical to the locus of the central point on the contact surface of a flat follower. The cam profile is drawn to touch the points of contact of construction lines representing the flat follower in various positions.

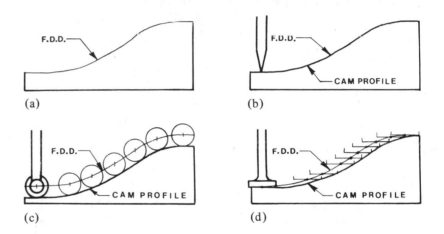

Fig. 4.11 Various shapes of wedge cam

It can be seen from fig. 4.11 that the profiles of wedge cams may be of varying shapes when transmitting identical motion, depending on the different types of follower used.

Example 1 – knife-edge follower

Cam specification: disc cam, minimum radius 40 mm; shaft diameter 35 mm; rotation anticlockwise

Displacement and motion:

    0°– 90°, rise of 30 mm with uniform acceleration
    90°–180°, rise of 30 mm with uniform retardation
    180°–240°, dwell (no motion or rest)
    240°–360°, fall with uniform velocity

To draw the cam profile (fig. 4.12):

1. Draw the follower displacement diagram, incorporating the cam minimum radius, which is the nearest approach of the follower to the centre of the cam.
2. Draw two circles, one of 40 mm cam minimum radius and the other of 100 mm cam maximum radius (i.e. the minimum cam radius plus the maximum follower displacement).
3. Draw radial lines at 30° intervals, numbering them in the reverse direction to the cam rotation. If extra accuracy is required, 15° or smaller intervals may be used as shown in the 300°–360° interval.
   Note: for construction purposes, the cam may be considered to be stationary while the follower rotates in the reverse direction about the cam.
4. Transfer all vertical distances from the follower displacement diagram to the corresponding radial lines, measuring from the centre.
5. With the help of a French curve, draw a smooth curve through the points obtained, to give the required cam profile.

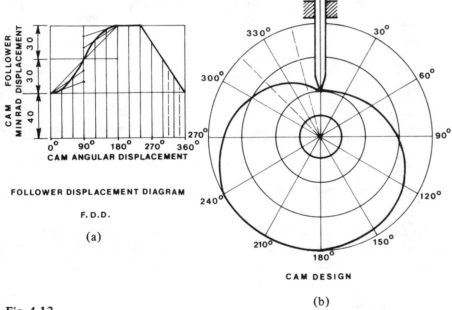

FOLLOWER DISPLACEMENT DIAGRAM

F.D.D.

(a)

CAM DESIGN

(b)

Fig. 4.12

Example 2 – roller follower

Cam specification: as example 1 (minimum radius is 40 mm minus the roller radius; i.e. 40 − 6 = 34 mm).
Displacement and motion: as example 1
Follower: 12 mm diameter roller

To draw the cam profile (fig. 4.13):

1. Draw the locus of the roller centre, which is an identical curve to the cam profile already constructed for the knife-edge follower in example 1 (fig. 4.12(b)).
2. With centres on this locus and radius 6 mm, draw a number of construction circles representing the roller in various positions.
3. Draw the best tangential curve to these circles to give the required cam profile.

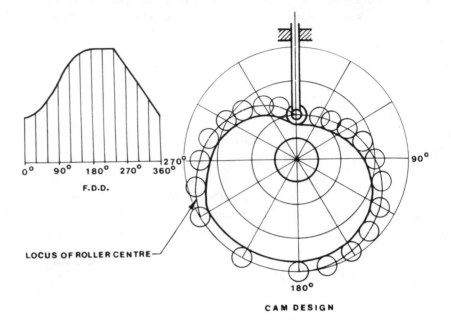

F.D.D.

LOCUS OF ROLLER CENTRE

CAM DESIGN

Fig. 4.13

Example 3 – flat follower

Cam specification: disc cam, minimum radius 46 mm; shaft diameter 38 mm; rotation clockwise
Displacement and motion:
    0°–180°, rise of 48 mm with s.h.m.
    180°–360°, fall of 48 mm with s.h.m.
Follower: flat, with 30 mm long contact surface

To draw the cam profile (fig. 4.14):

1. Draw the follower displacement diagram, F.D.D., incorporating the minimum cam radius.
2. Draw two circles, one of 46 mm cam minimum radius and another of 94 mm cam maximum radius.

18

3. Draw radial lines at 30° intervals, labelling them anticlockwise.
4. Transfer all the vertical distances from the follower displacement diagram to the corresponding radial lines of the cam, measuring from the centre.
5. At each of the plotted points on the radial lines, draw a line 30 mm long perpendicular to and bisected by the radial lines to represent the flat contact surface of the follower. These lines will be tangents to the cam profile.
6. Draw the best smooth curve to touch these tangents at the points of contact, remembering that a flat follower can be used only where the cam profile is convex.

To draw the cam profile (fig. 4.15):
1. Draw the base line of the follower displacement diagram, F.D.D., with vertical lines representing 30° intervals of rotation.
2. From point A on the base line, step off the 18 mm offset to obtain the point B. With centre B and radius 35 mm (the cam minimum radius) strike an arc intersecting the nearest vertical line at C.
3. From C draw a horizontal line, from which the follower curves of motion should be constructed in the usual way.
4. To draw the cam, with centre B draw a construction circle of 18 mm radius, equal to the amount of offset, and divide it with radial lines at 30° intervals.
   At the intersection points on the circumference, draw tangents to represent twelve positions of the follower.
5. With centre B draw a circle of 35 mm minimum radius, intersecting the vertical tangent at C.
6. Along this tangent, step off a distance CD equal to the 54 mm maximum follower rise. With centre B, draw a circle through the point D and label all intersection points in a clockwise direction.
7. Transfer all the vertical distances from the follower displacement diagram to the corresponding tangents, measuring from the points of tangency.
8. With the help of a French curve, draw a smooth curve through the points obtained in step 7. This is the required cam profile.

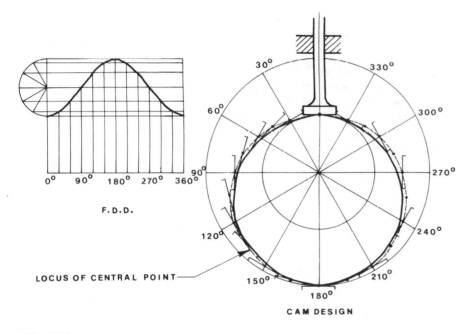

F.D.D.

LOCUS OF CENTRAL POINT

CAM DESIGN

Fig. 4.14

Example 4 – offset knife-edge follower
Cam specification: disc cam, minimum radius 35 mm; rotation anticlockwise
Displacement and motion:
   0°–150°, 54 mm rise with uniform velocity
   150°–180°, dwell
   180°–360°, 54 mm fall with s.h.m.
Follower: knife-edge, offset 18 mm to the right of the cam centre line

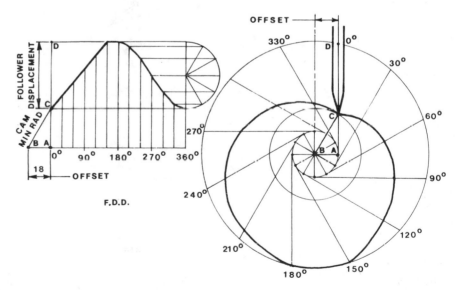

F.D.D.

CAM DESIGN

Fig. 4.15

19

Example 5 – offset roller follower

Cam specification: as example 4
Displacement and motion: as example 4
Follower: 16 mm diameter roller offset 18 mm to the right of the cam centre line

To draw the cam profile (fig. 4.16):
1. Draw the locus of the roller centre, which is an identical curve to the cam profile already constructed for the knife-edge follower in example 4 (fig. 4.15).
2. With centres on this locus, draw a number of construction circles of 16 mm diameter to represent the roller in various positions.
3. Draw the best tangential curve to these circles to give the required cam-profile curve.

Example 6 – face cam

Cam specification: face cam; minimum cam and roller centre distance 28 mm; shaft diameter 20 mm; rotation clockwise
Displacement and motion:
  0°–180°, 36 mm rise with s.h.m.
180°–360°, 36 mm fall with s.h.m.
Follower: 18 mm diameter roller

To draw the cam groove (fig. 4.17):
1. Draw the follower displacement diagram, F.D.D., incorporating the minimum cam radius.
2. Draw a construction circle of 80 mm cam maximum radius.
3. Draw radial lines at 30° intervals and label them in an anticlockwise direction.
4. Transfer all the vertical distances from the follower displacement diagram to the corresponding radial lines, measuring from the centre.
5. Draw a smooth curve through the points obtained to give the locus of the centre of the follower and, with centres on this locus, draw a number of construction circles of 18 mm diameter to represent the roller follower in various positions.
6. Draw the two best tangential curves touching the construction rollers. These form the required cam groove.

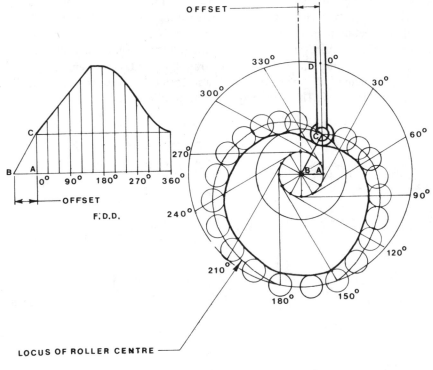

Fig. 4.16

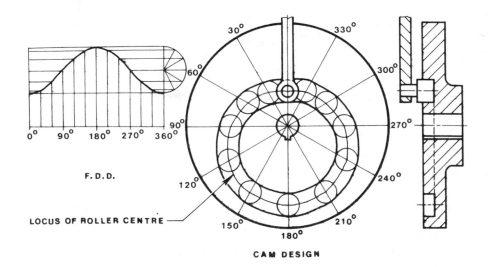

Fig. 4.17

20

Example 7 – radial-arm roller follower

Cam specification: disc cam, minimum radius 20 mm; shaft diameter 22 mm; rotation clockwise

Displacement and motion:

$0°-120°$, 36 mm rise with uniform velocity
$120°-240°$, dwell
$240°-360°$, 36 mm fall with uniform velocity
Follower: radial arm, 110 mm long, with a 16 mm diameter roller follower, positioned on the cam centre line at 110 mm to the left of the cam centre.

To draw the cam profile (fig. 4.18):

1. Draw the required follower displacement diagram, F.D.D.
2. Position the pivot centre A_1 relative to the cam centre O as shown.
3. With centre O, draw a circle of 110 mm radius through the pivot centre A_1 and draw radial intersection points A_2, A_3, . . . , A_{12} at $30°$ intervals representing the pivot at twelve positions of the cam.
4. With centre O, draw a circle of 28 mm (20 mm cam minimum radius plus 8 mm roller radius) representing the lowest position of the roller centre.
5. With centre O draw a circle of 64 mm (28 mm plus the maximum follower rise of 36 mm) representing the highest position of the roller centre.
6. With centre A_1 and radius A_1O (110 mm), draw an arc to intersect the lowest and highest position of the roller-centre circles at B_1 and C_1 respectively.
   Repeat this procedure with centre A_2 to intersect at B_2 and C_2 etc.

7. With centre O and radii equal to the corresponding vertical distances on the follower displacement diagram, strike arcs on the BC curves to obtain the various positions of the roller centres.
8. Draw circles of 16 mm roller diameter at these positions.
9. Draw the best cam-profile curve tangential to these circles.

Note. If the pivot was positioned to the right of the cam and the cam rotation was anticlockwise, the construction would be the mirror image of that in fig. 4.18, as shown in fig. 4.19.

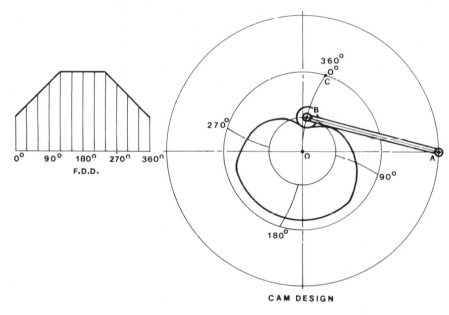

Fig. 4.19

CAM DESIGN

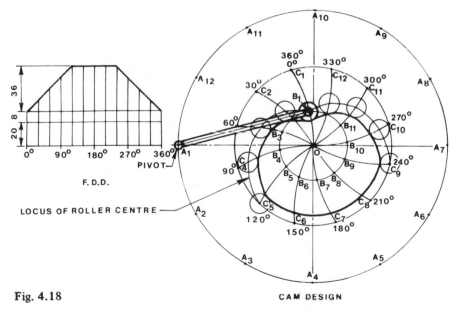

Fig. 4.18

CAM DESIGN

4.5 Test questions

1. Construct any of the cams shown in example 1 page 17 to example 7 page 21.
2. Define a cam and a follower.
3. Name and describe the five main types of cam.
4. Sketch and name three types of cam follower.
5. Sketch a cam and an offset follower to give an oscillating motion.

21

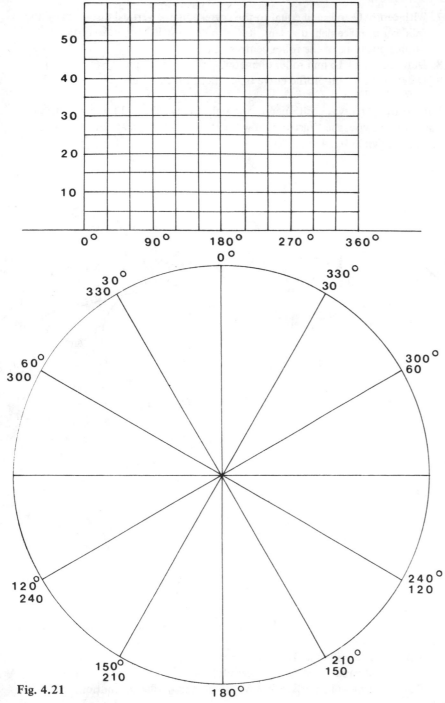

Fig. 4.21

6. Figure 4.20 shows a cam-follower displacement diagram with three different follower-motion curves. Identify each of these curves.

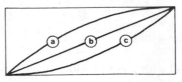

Fig. 4.20

7. Construct three cam-follower displacement diagrams between 0° and 180° to give a rise of 60 mm and the following motions:
a) uniform velocity,
b) uniform acceleration and retardation,
c) simple harmonic motion.
(Tracing paper and fig. 4.21 may be used.)

8. Construct a cam to satisfy the following design specifications: a disc cam of minimum radius 40 mm mounted on a camshaft of 40 mm diameter to rotate in an anticlockwise direction with a knife-edge follower to have the following displacement and motion:
   0°–180°, rise of 70 mm with simple harmonic motion;
180°–360°, fall of 70 mm with uniform velocity.
(You may draw to half full-size scale, using tracing paper and fig. 4.21.)

9. Using tracing paper and fig. 4.21, construct a cam of minimum radius 15 mm to rotate in an anticlockwise direction. The cam-follower displacement and motion is to be a rise of 40 mm with simple harmonic motion for half a revolution and a fall with simple harmonic motion for the remaining half a revolution. Construct and compare superimposed cam profiles for the following followers: (a) knife-edge, (b) flat with 20 mm long contact surface.

10. Construct a cam to satisfy the following design specifications: a disc cam of a minimum radius 40 mm mounted on a camshaft of 24 mm diameter to rotate in a clockwise direction; a cam follower with a roller of 20 mm diameter to have the following displacement and motion:
   0°–120°, rise of 30 mm with uniform acceleration;
120°–240°, rise of 30 mm with uniform retardation;
240°–360°, fall of 60 mm with uniform velocity.
(You may draw to half full-size scale, using tracing paper and fig. 4.21.)

11. Construct a cam of minimum radius 30 mm mounted on a camshaft of 30 mm diameter to rotate in an anticlockwise direction; a flat cam follower with 24 mm long contact surface to have the following displacement and motion:
   0°–150°, rise of 80 mm with uniform velocity;
150°–180°, dwell;
180°–360°, fall of 80 mm with simple harmonic motion.
(You may draw to half full-size scale, using tracing paper and fig. 4.21.)

5 Gears

Motion can be transmitted from one parallel shaft to another by means of friction cylinders, as shown in fig. 5.1(a).

Such arrangements of cylinders rolling tightly together are suitable only for low-torque applications. To eliminate the possibility of slip between the friction cylinders and to enable a high torque to be transmitted, a positive drive is obtained by cutting recesses and adding projections to the cylinders as shown in fig. 5.1(b). These 'cylinders' with teeth so formed are called *spur gears*, and the circles corresponding to the original reference friction cylinders are called the reference or pitch circles.

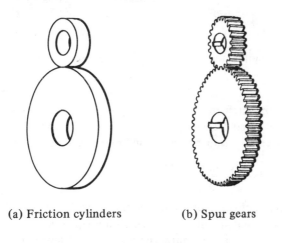

(a) Friction cylinders   (b) Spur gears

Fig. 5.1

In order to keep the movement ratio of the two mating (meshing) gears constant (i.e. a given movement of the driver gear always produces the same movement of the driven gear) and to ensure a rolling motion without slip between two gear teeth in contact, the profile of the teeth is of either cycloidal or involute form (see pages 2 and 3).

The involute profile is most commonly used, as it has the following advantages:
a) it is easily manufactured and measured;
b) the path of the tooth contact point is a straight line, instead of a curve as for cycloidal teeth; and
c) involute teeth are usually stronger than the corresponding cycloidal teeth.

5.1 Types of gear
There are three main types of gear: spur gears, bevel gears, and worm-and-wheel gears.

Spur gears
These gears are used for transmitting motion between two parallel shafts, and a spur gear has its teeth cut parallel to the axis of its shaft as shown in fig. 5.1(b). The conventional representation is shown in fig. 5.2.

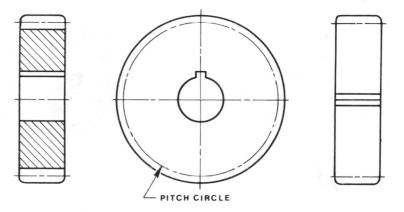

PITCH CIRCLE

Fig. 5.2 Conventional representation of spur gears

Helical gears act similarly to spur gears but their teeth are cut along helices. The point of contact between mating teeth moves steadily across the tooth flanks instead of there being a sudden engagement for a whole tooth width as is the case with spur gears. Helical gears run quietly and smoothly and are wear-resistant.

When a single helical gear is used, an axial thrust which tends to separate the gears is produced, as shown in fig. 5.3(a). To counteract this, double-helical gears can be used, as shown in fig. 5.3(b).

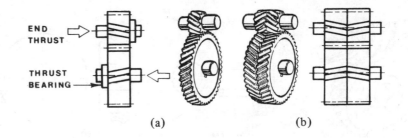

END THRUST

THRUST BEARING

(a)   (b)

Fig. 5.3 Helical gears: (a) single, (b) double

Bevel gears (fig. 5.4(b))

These gears are used for transmitting motion between shafts whose axes intersect. The action of bevel gears may be compared with two friction cones rolling tightly together. The friction cones represent the reference or pitch cones of the bevel gears and their base diameters are pitch-circle diameters, d_w, as shown in fig. 5.4(a). The cone distance, R, is the slant height of the pitch cone and must be the same for both mating gears. When the gear teeth are manufactured, they are progressively smaller towards the cone apex.

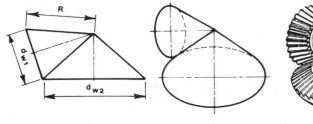

(a) Friction cones   (b) Bevel gears

Fig. 5.4 Bevel gears

Worm-and-wheel gears

These gears are used for transmitting motion between two non-intersecting, non-parallel shafts whose axes are usually perpendicular. A worm, which may be considered to be a single- or multi-start thread screw, drives a worm wheel which has a larger pitch-circle diameter, as shown in fig. 5.5.

The teeth on the worm wheel are throated (radiused) to give greater contact with the worm, to ensure effective mating. The axial pitch of the worm is equivalent to the circular pitch of the worm wheel.

The gear ratio — i.e. the number of teeth on the wheel to the number of threads on the worm — is usually very high.

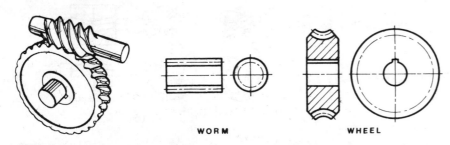

WORM   WHEEL

Fig. 5.5 Conventional representation of a worm-and-wheel gear

5.2 Spur-gear proportions (see basic-formulae note, page 26)

When gear teeth are formed on pitch cylinders or circles, the 'added' projection is called the *addendum* (h_a), and it terminates at the tip of a tooth. The 'deducted' recess is called the *dedendum* (h_f) and it terminates at the root of a tooth, as shown in fig. 5.6(a).

The dedendum is one-and-a-quarter times as long as the addendum. This additional length of a quarter addendum provides the necessary clearance (c) between the tips of the mating teeth of a gear pair, as shown in fig. 5.6(b).

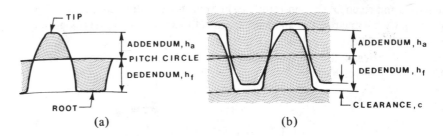

Fig. 5.6 (a) Addendum and dedendum, (b) clearance

The ratio of the pitch-circle diameter (d_w) to the total number of teeth (z) is very important when designing gears. It is called the module (m) and, like the pitch-circle diameter, it is measured in millimetres as shown in fig. 5.7(a). Hence

$$\text{module } m = \frac{\text{pitch-circle diameter}}{\text{number of teeth}} = \frac{d_w}{z} \text{ mm}$$

Also, the length of the addendum is always equal to the module in millimetres ($h_a = m$ mm), and consequently the dedendum is equal to one-and-a-quarter times the module ($h_f = 1.25m$ mm). The module must always be the same for all mating gears.

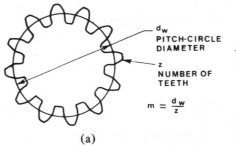

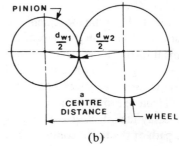

Fig. 5.7 (a) Module, (b) centre distance

The centre distance (a) between any two mating gears is the sum of their pitch-circle radii, as shown in fig. 5.7(b). The smaller of two mating gears is called the *pinion* and the larger is called the *wheel*.

The *tip* and *root circles* contain the tips and roots of the gear teeth, and the corresponding diameters are called the *tip-* and *root-circle diameters*.

Example   Two parallel shafts are to be connected by a spur gear wheel and pinion. The wheel has 18 teeth and a module of 10 mm. If the gear ratio, which is the ratio of the number of teeth on the wheel to the number of teeth on the pinion, is 3:2, calculate (a) the addendum (h_a), (b) the dedendum (h_f), (c) the centre distance (a), (d) the tip-circle diameter of the pinion (d_{a1}), (e) the root-circle diameter of the wheel (d_{f2}).

a)   Addendum = module = 10 mm

b)   Dedendum = 1.25 × module = 12.5 mm

c)   Gear ratio (u) = $\dfrac{\text{number of teeth of wheel } (z_2)}{\text{number of teeth of pinion } (z_1)}$

∴   number of teeth of pinion (z_1) = $\dfrac{\text{number of teeth of wheel } (z_2)}{\text{gear ratio } (u)}$

$$= \frac{18}{3/2} = 12 \text{ teeth}$$

Pinion pitch-circle diameter = module (m) × number of teeth of pinion (z_1)

$$= 10 \text{ mm} \times 12 = 120 \text{ mm}$$

Wheel pitch-circle diameter = module (m) × number of teeth of wheel (z_2)

$$= 10 \text{ mm} \times 18 = 180 \text{ mm}$$

Centre distance (a) = pinion pitch-circle radius + wheel pitch-circle radius

$$= (120 \text{ mm})/2 + (180 \text{ mm})/2 = 150 \text{ mm}$$

d)   Pinion tip-circle diameter (d_a) = pinion pitch diameter (d_w) + 2 × addendum (h_a)

$$= 120 \text{ mm} + 2 \times 10 \text{ mm} = 140 \text{ mm}$$

e)   Wheel root-circle diameter (d_f) = wheel pitch diameter (d_w) − 2 × dedendum (h_f)

$$= 180 \text{ mm} - 2 \times 12.5 \text{ mm} = 155 \text{ mm}$$

5.3 General definitions (figs 5.8 and 5.9) (see basic formulae, page 26)
A *gear tooth* is a projecting part of a gear and is designed to turn the other gear by means of contact of mating teeth.

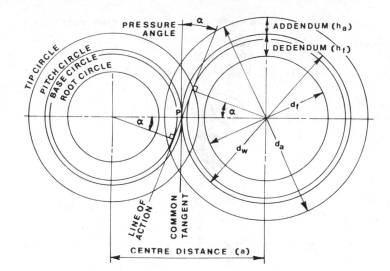

Fig. 5.8 Gear circles and angles

A *gear pair* is a mechanism consisting of two mating gears, one of which turns the other by the action of teeth which are successively in contact.

The *pinion* is the smaller and the *wheel* is the larger of the gear pair.

The *gear ratio* (u) is the ratio of the number of teeth on the wheel to the number of teeth on the pinion.

The *pitch circle* represents the friction cylinder or cone which would transmit the motion between two mating gears without slip and its diameter is the pitch-circle diameter (d_w).

The *pitch point* (P) is the point of contact between the pitch circles of two mating gears.

The *centre distance* (a) is the shortest distance between the axes of a gear pair and it consists of the sum of the pitch-circle radii of the gears.

The *base circle* is the imaginary circle from which the involute is generated.

The *tip circle* is the circle containing the tips or crests of the teeth and its diameter is the *tip-circle diameter* (d_a).

The *root circle* is the circle containing the roots or bases of the tooth spaces between adjacent teeth and its diameter is the root-circle diameter (d_f).

The *line of action* is the line along which the contact takes place between the mating teeth. It passes through the pitch point and is tangential to the base circles of the gear pair.

The *pressure angle* (α) is the angle between the line of action and the common tangent to the pitch circles and is usually 20°.

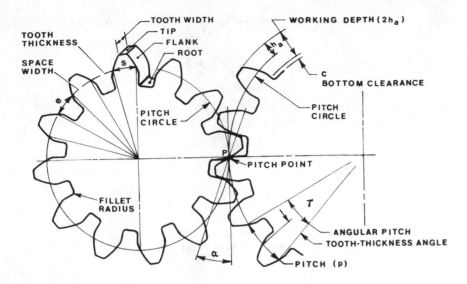

Fig. 5.9 Involute-gear definitions

The *addendum* (h_a) is the radial distance between the tip circle and the pitch circle.

The *dedendum* (h_f) is the radial distance between the root circle and the pitch circle.

The *working depth* is the distance between the tips of two mating teeth measured along the line of centres.

The *tooth flank* is the tooth profile lying between the tip and the root of a tooth.

The *bottom clearance* (c) is the distance between the root of a gear and the tip of the mating gear, measured along the line of centres.

The *space width* (e) is the length of the pitch-circle arc between two adjacent teeth.

The *tooth thickness* (s) is the length of the pitch-circle arc between the two flanks of a tooth and it equals the space width.

The *tooth-thickness angle* is the angle subtended by the tooth thickness at the centre of the gear.

The *(transverse) pitch* (p) is the length of the pitch-circle arc lying between two consecutive corresponding profiles and it equals the sum of the tooth thickness and the space width.

The *angular pitch* (τ) is the ratio of the angle subtended by the circumference to the number of teeth (z) and is expressed in degrees: $\tau = 360°/z$.

The *module* (m) is the ratio of the pitch-circle diameter to the number of teeth and is expressed in millimetres.

5.4 Basic formulae

Note: subscript f refers to root measurements,
a refers to tip measurements,
w refers to pitch-circle measurements,
b refers to base-circle measurements,
1 refers to pinion measurements,
2 refers to wheel measurements.

$$\text{Module } m = \frac{\text{pitch-circle diameter (mm)}}{\text{number of teeth}} = \frac{\text{P.C.D.}}{z}$$

$$m = \frac{d_w}{z} \text{ mm}$$

$$\text{Pitch } p = \frac{\text{circumference of pitch circle}}{\text{number of teeth}}$$

$$p = \frac{\pi \times d_w}{z} = \pi \times m$$

Tooth thickness s = half of the pitch = $p/2$

Addendum h_a = module = m

Clearance c = 0.25 × addendum = 0.25 × h_a

Dedendum h_f = addendum + clearance = $h_a + c$

$$= h_a + 0.25 \times h_a = 1.25 h_a = 1.25 m$$

$$\text{Gear ratio } u = \frac{\text{number of teeth in wheel}}{\text{number of teeth in pinion}} = \frac{z_2}{z_1}$$

$$= \frac{\text{wheel pitch-circle diameter}}{\text{pinion pitch-circle diameter}} = \frac{d_{w2}}{d_{w1}}$$

$$\text{Centre distance } a = \frac{\text{wheel pitch diameter + pinion pitch diameter}}{2}$$

$$= \frac{d_{w2} + d_{w1}}{2}$$

$$\text{Angular pitch } \tau = \frac{\text{angle subtended by circumference}}{\text{number of teeth}} = \frac{360°}{z}$$

$$\begin{array}{l}\text{Base-circle} \\ \text{diameter}\end{array} = \begin{array}{l}\text{pitch-circle} \\ \text{diameter}\end{array} \times \begin{array}{l}\text{cosine of the} \\ \text{pressure angle}\end{array}$$

$$d_b = d_w \times \cos \alpha$$

Fillet radius = 0.4 × module

Note: The module, pitch, tooth thickness, addendum, clearance, and dedendum of each of the mating gears — pinion and wheel — must always be the same.

5.5 Construction of an involute-tooth profile

The method of construction of the involute curve was discussed on page 3. Figure 5.10(a) shows a portion of the involute curve generated from the base circle. This is sufficient in length for the required profile of a tooth. The length of each tangent is equal to the length of the base-circle arc measured from point a to the corresponding point of tangency.

Construction of an approximate involute-tooth profile

It is very tedious and time-consuming to draw accurate gear-teeth profiles, and approximate involute curves represented by drawing circular arcs are often adequate.

To draw an approximate involute-tooth profile (fig. 5.10(b)):
1. Draw the tip, base, and root circles with centre O.
2. Along a radial line, mark a point A on the tip circle and point B on the base circle. Divide AB into *three* equal parts to obtain point C such that $AC = \frac{1}{3}AB$.
3. From C, draw CD tangential to the base circle with D being the point of tangency. Divide CD into *four* equal parts to obtain point E such that $ED = \frac{1}{4}CD$.
4. With centre E, strike an arc to start at the base circle, pass through C, and end at the tip circle. This arc is the approximation to the involute gear-tooth profile.
5. From the point at which this curve intersects the base circle, draw a radial line towards the centre O and blend it with the root circle by drawing a fillet of radius equal to approximately four-tenths of the module, $0.4m$.
6. With centre O, draw through E a circle which is to be used for the centres of all the tooth-profile arcs.

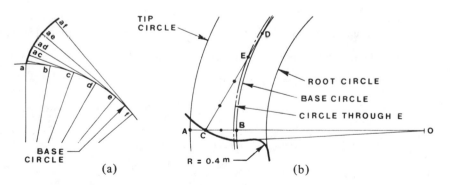
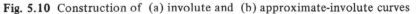

Fig. 5.10 Construction of (a) involute and (b) approximate-involute curves

7. With centres on the circle through E and radius EC, draw the rest of the arcs, ensuring that they cut the pitch circle at the appropriate points, and complete the required tooth profiles as in step 5 (see fig. 5.12).

5.6 Spur-gear calculations

Example 1   Spur gears are to be used to transmit power from one shaft to another shaft parallel to it. The gear ratio (u) is 4:1 and the pinion has 20 teeth and revolves at 2 rev/s. The module for both gears is 10 mm. Determine (a) the speed of the gear wheel in rev/min, (b) the pitch-circle diameters, (c) the tip diameters, (d) the root diameters, (e) the circular pitch, (f) the circular tooth thickness, (g) the centre distance of the mating (meshing) gears.

a) The speeds of gears are inversely proportional to their respective number of teeth:

$$\frac{\text{speed of pinion (rev/s)}}{\text{speed of wheel (rev/s)}} = \frac{\text{number of teeth of wheel}}{\text{number of teeth of pinion}}$$

or $\qquad\qquad \dfrac{n_1}{n_2} = \dfrac{z_2}{z_1}$

where subscript 1 refers to the pinion and subscript 2 refers to the wheel.

But z_2/z_1 = gear ratio u

$\therefore$ speed of wheel, $n_2 = \dfrac{z_1}{z_2} \times n_1 = \dfrac{1}{\text{gear ratio}} \times n_1$

$$= \tfrac{1}{4} \times 2 \text{ rev/s} = \tfrac{1}{4} \times 2 \times 60 \text{ rev/min} = 30 \text{ rev/min}$$

b)   Gear ratio $u = z_2/z_1 = 4/1$

$\therefore$ z_2 = gear ratio $\times z_1 = (4/1) \times 20 = 80$

i.e. the gear wheel has 80 teeth.

Now   module $m = \dfrac{\text{pitch-circle diameter}}{\text{number of teeth}} = \dfrac{d_w}{z}$

$\therefore$ pinion pitch-circle diameter, $d_{w1} = m \times z_1 = 10 \text{ mm} \times 20 = 200 \text{ mm}$

and   wheel pitch-circle diameter, $d_{w2} = m \times z_2 = 10 \text{ mm} \times 80 = 800 \text{ mm}$

c)   Tip diameter = pitch-circle diameter + 2 $\times$ addendum

$$d_a = d_w + 2h_a$$

where $h_a = m = 10 \text{ mm}$

$\therefore$ pinion tip diameter d_{a1} $= d_{w1} + 2h_a$

$$= 200 \text{ mm} + 2 \times 10 \text{ mm} = 220 \text{ mm}$$

and wheel tip diameter d_{a2} $= d_{w2} + 2h_a$

$$= 800 \text{ mm} + 2 \times 10 \text{ mm} = 820 \text{ mm}$$

d) Root diameter = pitch-circle diameter $-$ 2 x dedendum

$$d_f = d_w - 2h_f$$

But dedendum h_f = addendum + clearance

$$= h_a + 0.25h_a = 1.25h_a$$

$$= 1.25 \times 10 \text{ mm} = 12.5 \text{ mm}$$

$\therefore$ pinion root diameter d_{f1} $= d_{w1} - 2h_f$

$$= 200 \text{ mm} - 2 \times 12.5 \text{ mm} = 175 \text{ mm}$$

and wheel root diameter d_{f2} $= d_{w2} - 2h_f$

$$= 800 \text{ mm} - 2 \times 12.5 \text{ mm} = 775 \text{ mm}$$

e) Circular pitch $p = \dfrac{\text{circumference of pitch circle}}{\text{number of teeth}} = \dfrac{\pi d_w}{z}$

But $d_w/z = m$

$\therefore$ $p = \pi m$

$\therefore$ $p_1 = p_2 = \pi \times 10 \text{ mm} = 31.42 \text{ mm}$

f) Circular tooth thickness $= \frac{1}{2} \times$ circular pitch

$\therefore$ $s_1 = s_2 = p/2$

$$= (31.42 \text{ mm})/2 = 15.71 \text{ mm}$$

g) The centre distance is equal to the sum of the pitch-circle radii of the mating gears,

$\therefore$ $a = \dfrac{d_{w1}}{2} + \dfrac{d_{w2}}{2}$

$$= \dfrac{200 \text{ mm}}{2} + \dfrac{800 \text{ mm}}{2} = 500 \text{ mm}$$

Example 2 The maximum centre distance between two parallel spur gears is not to exceed 220 mm. The gear ratio is to be 4:3 with a module of 12 mm and a pressure angle of 20°. Do all the calculations necessary for the design of the gear pair.

For a given size of gear tooth, the number of teeth on a gear is proportional to the pitch-circle diameter and hence to the pitch-circle radius of the gear.

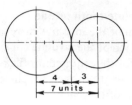

Fig. 5.11

Figure 5.11 shows two gears with pitch-circle radii in the ratio 4:3. Let the centre distance a, which equals seven units of radius, be 220 mm; then

approximate wheel pitch-circle radius $= \frac{4}{7} \times 220 \text{ mm} = 125.7 \text{ mm}$

$\therefore$ $d_{w2} = 2 \times 125.7 \text{ mm} = 251.4 \text{ mm}$

and approximate pinion pitch-circle radius $= \frac{3}{7} \times 220 \text{ mm} = 94.3 \text{ mm}$

$\therefore$ $d_{w1} = 2 \times 94.3 \text{ mm} = 188.6 \text{ mm}$

approximate pinion number of teeth, $z_1 = \dfrac{d_{w1}}{m} = \dfrac{188.6 \text{ mm}}{12 \text{ mm}} = 15.71$

and approximate wheel number of teeth, $z_2 = \dfrac{d_{w2}}{m} = \dfrac{251.4 \text{ mm}}{12 \text{ mm}} = 20.95$

Based on these approximate figures, we can now calculate the actual design features as follows.

a) There must be a whole number of teeth on the wheel and pinion, therefore, from the calculations above, for a gear ratio of 4:3 the most appropriate number-of-teeth ratio will be 20:15; i.e. $z_1 = 15$ and $z_2 = 20$.

b) Actual pinion pitch-circle diameter $d_{w1} = z_1 m = 15 \times 12 \text{ mm} = 180 \text{ mm}$

and actual wheel pitch-circle diameter $d_{w2} = z_2 m = 20 \times 12 \text{ mm} = 240 \text{ mm}$

c) Actual centre distance = sum of pitch-circle radii

$$= \dfrac{d_{w1}}{2} + \dfrac{d_{w2}}{2} = \dfrac{180 \text{ mm}}{2} + \dfrac{240 \text{ mm}}{2} = 210 \text{ mm}$$

d) Tip diameter = pitch-circle diameter + 2 x addendum

$$d_a = d_w + 2h_a$$

where $h_a = m = 12 \text{ mm}$

∴　　pinion tip diameter d_{a1} $= d_{w1} + 2h_a$

$\qquad\qquad\qquad = 180 \text{ mm} + 2 \times 12 \text{ mm} = 204 \text{ mm}$

and　wheel tip diameter d_{a2} $= d_{w2} + 2h_a$

$\qquad\qquad\qquad = 240 \text{ mm} + 2 \times 12 \text{ mm} = 264 \text{ mm}$

e)　Root diameter = pitch-circle diameter − 2 × dedendum

$$d_f = d_w - 2h_f$$

where　$h_f = 1.25 h_a$ (see example 1(d))

∴　　pinion root diameter d_{f1} $= d_{w1} - 2h_f$

$\qquad\qquad\qquad = 180 \text{ mm} - 2 \times 1.25 \times 12 \text{ mm} = 150 \text{ mm}$

and　wheel root diameter d_{f2} $= d_{w2} - 2h_f$

$\qquad\qquad\qquad = 240 \text{ mm} - 2 \times 1.25 \times 12 \text{ mm} = 210 \text{ mm}$

f)　Circular pitch $p = \dfrac{\text{circumference of pitch circle}}{\text{number of teeth}}$

$$= \frac{\pi d_w}{z}$$

$$= \pi m \quad (\text{since } d_w/z = m)$$

∴　　$p = \pi \times 12 \text{ mm} = 37.7 \text{ mm}$

g)　Angular pitch $\tau = \dfrac{\text{angle subtended by circumference}}{\text{number of teeth}}$

∴　　$\tau_1 = \dfrac{360°}{z_1} = \dfrac{360°}{15} = 24°$

and　$\tau_2 = \dfrac{360°}{z_2} = \dfrac{360°}{20} = 18°$

h)　Tooth-thickness angle $= \frac{1}{2} \times$ angular pitch

∴　　$s_1 = 24°/2 = 12°$

and　$s_2 = 18°/2 = 9°$

j)　$\begin{array}{l}\text{Base-circle} \\ \text{diameter}\end{array} = \begin{array}{l}\text{pitch-circle} \\ \text{diameter}\end{array} \times \begin{array}{l}\text{cosine of the} \\ \text{pressure angle}\end{array}$

$$d_b = d_w \times \cos \alpha$$

∴　　pinion base-circle diameter d_{b1} $= d_{w1} \times \cos \alpha$

$\qquad\qquad\qquad = 180 \text{ mm} \times \cos 20° = 169.1 \text{ mm}$

and　wheel base-circle diameter d_{b2} $= d_{w2} \times \cos \alpha$

$\qquad\qquad\qquad = 240 \text{ mm} \times \cos 20° = 225.5 \text{ mm}$

k)　Fillet radius = 0.4 × module

$\qquad\qquad\qquad = 0.4 \times 12 \text{ mm} = 4.8 \text{ mm}$

Example 3　Draw a spur gear pair using the following data: $m = 16$ mm, $z_1 = 12$, and $z_2 = 18$.

We must first calculate the other design features, as follows.

The pitch-circle diameters are

$$d_{w1} = m \times z_1 = 16 \text{ mm} \times 12 = 192 \text{ mm}$$

and　$d_{w2} = m \times z_2 = 16 \text{ mm} \times 18 = 288 \text{ mm}$

The tip diameters are

$$d_{a1} = d_{w1} + 2h_a = 192 \text{ mm} + 2 \times 16 \text{ mm} = 224 \text{ mm}$$

and　$d_{a2} = d_{w2} + 2h_a = 288 \text{ mm} + 2 \times 16 \text{ mm} = 320 \text{ mm}$

The root diameters are

$$d_{f1} = d_{w1} - 2h_f = 192 \text{ mm} - 2 \times 1.25 \times 16 \text{ mm} = 152 \text{ mm}$$

and　$d_{f2} = d_{w2} - 2h_f = 288 \text{ mm} - 2 \times 1.25 \times 16 \text{ mm} = 248 \text{ mm}$

Pinion angular pitch $= \dfrac{360°}{z_1} = \dfrac{360°}{12} = 30°$

Wheel angular pitch $= \dfrac{360°}{z_2} = \dfrac{360°}{18} = 20°$

The construction now proceeds as follows (fig. 5.12):

1. Draw the two tangential pitch circles of radii (192 mm)/2 = 96 mm and (288 mm)/2 = 144 mm.
2. At the pitch point, draw the common tangent and the line of action inclined to the common tangent at 20° and passing through the pitch point at P.
3. Tangential to the line of action, draw the base circles. These circles are to be used *only for generation of the involute curves*.
4. Draw the tip circles with radii of (224 mm)/2 = 112 mm and (320 mm)/2 = 160 mm and the root circles with radii of (152 mm)/2 = 76 mm and (248 mm)/2 = 124 mm.

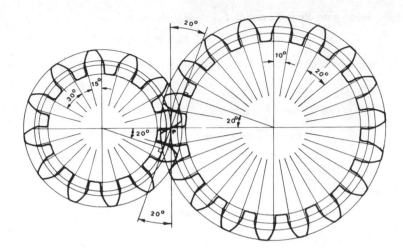

Fig. 5.12 Construction of involute-tooth profiles

All points of contact between the mating gear teeth will lie between the base circles and the tip circles. There is no contact between teeth between the base circles and the root circles, where undercuts are sometimes machined for clearance purposes.

5. Starting from the pitch point, mark off points which divide the pinion pitch circle into arcs subtending the angular pitch increment of 30° and bisect these arcs to obtain the angular tooth thickness of 15° for the pinion.

    Similarly obtain the angular tooth thickness of 10° for the wheel.

    Note: Each generated involute curve must start at the base circle, pass through one of the points marked on the pitch circle, and end at the tip circle of the gear.

6. Generate the involute curve from the pinion base circle (see fig. 5.10(a), page 27).

7. Place tracing paper over the generated curve and locate it by means of a compass pin at the centre of the pinion. On both sides of the tracing paper, very carefully copy the involute curve with a soft pencil.

    Rotate the tracing paper about the pinion centre and trace an involute curve passing through each intersection point on the pitch circle, thus obtaining one side of the required tooth profiles. Reverse the tracing paper to obtain the other side of the tooth profiles.

8. To complete the drawing, repeat stages 6 and 7 for the wheel, generating the involute curve from the wheel base circle.

    Note: The tooth profiles are usually drawn using the approximate method indicated in fig. 5.10(b), page 27.

5.7 The involute rack

The involute rack may be regarded as a small part of an involute gear of infinitely large pitch-circle diameter, such that the pitch-circle arc becomes a pitch line. Also, the involute of an infinitely large circle is a straight line which forms the flanks of the rack teeth that are straight and normal (at 90°) to the line of action, as shown in fig. 5.13.

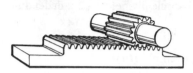

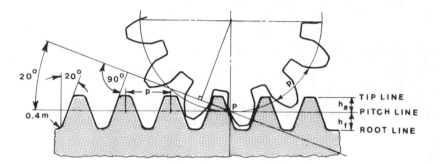

Fig. 5.13 Involute rack

For the rotating pinion to impart a straight-line motion to the rack and to engage correctly, both the rack and the pinion must have the same module, addendum, dedendum, pitch, and tooth thickness.

Example Draw an involute rack using the following data: $m = 16$ mm and $z_1 = 12$.

To draw an approximate involute rack to engage (mesh) with a pinion:

1. Draw the rack pitch line tangential to the pinion pitch circle. Then draw the tip line parallel to the pitch line and at a distance from it equal to the addendum (m). Similarly draw the root line at a distance equal to the dedendum ($1.25m$).

2. Starting from the common pitch point, mark off on the pitch line distances equivalent to the circular pitch of the pinion ($p = \pi m$) and bisect them to obtain the required tooth thicknesses.

3. Through these points, draw the tooth flanks inclined at 90° to the line of action and blend them with the root line by drawing the fillet radii of approximately $0.4m$.

5.8 Bevel gears

Usually only about one third of the cone distance is used as the width of the toothed part of the bevel gear. It is measured along the pitch-cone surface and is called the face width, b, as shown in fig. 5.14(a).

The common apex is the point where two pitch-cone axes intersect, and the angle between these axes is called the shaft angle, Σ. The angle between the pitch-cone surface and the axis is called the pitch angle, δ_w, as shown in fig. 5.14(a).

The conventional representation of bevel gears is shown in fig. 5.14(b).

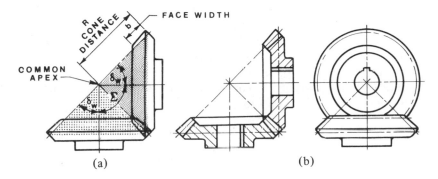

Fig. 5.14 (a) Pitch and shaft angles and (b) conventional representation of bevel gears

A bevel gear with a pitch angle of 90° is called a *crown gear* or *crown wheel*, and one with a pitch angle of 45° is called a *mitre*, as shown in fig. 5.15.

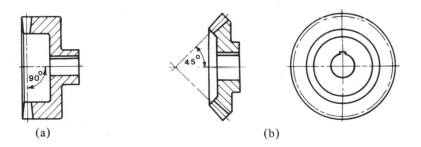

Fig. 5.15 (a) Crown gear and (b) mitre

Manufactured bevel gears include the additional back cone whose surface is perpendicular to the surface of the pitch cone. When viewed on the back cone, bevel gear teeth are identical with spur-gear involute teeth and are generated in the same way, as shown in fig. 5.16.

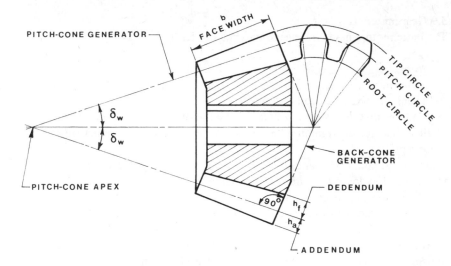

Fig. 5.16 Geometry of bevel gears

The addendum and dedendum are measured radially along the back-cone surface, as shown in fig. 5.16.

The pitch, tip, and root circles and the corresponding diameters are measured in a plane perpendicular to the axis, as shown in fig. 5.17.

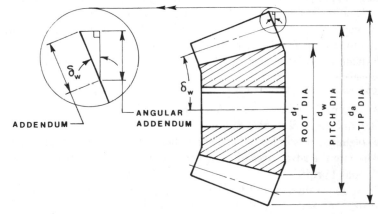

Fig. 5.17 Bevel-gear geometry

To calculate the tip and root diameters, the angular addendum and angular dedendum must be added and subtracted respectively, as for spur gears. (Angular addendum = addendum x cosine of pitch angle δ_w)

31

5.9 Gear materials

The basic requirements for a gear material are that it should be

a) rigid,
b) strong,
c) hard-wearing, and
d) capable of being shaped accurately.

A harder material is usually used for the pinion than the wheel, as, due to its smaller diameter, the pinion revolves more often than the wheel and thus wears more rapidly.

The most common gear materials are carbon and alloy steels in their various forms. For low loads the carbon content should not be less than 0.4%, but for high loads the steel is usually case-hardened after the teeth have been cut.

Low-load and low-speed gears are often made of cast iron, bronzes, and plastics materials.

Worm wheels are made of various bronzes, the worms being made of steels which are hardened and ground to obtain the accurate and smooth surfaces required for quiet running and long life.

5.10 Test questions

1. Define spur gears.
2. What are the advantages of using helical gears?
3. Define bevel gears.
4. Define worm-and-wheel gears.
5. Define (a) a gear pair; (b) the gear ratio; (c) the centre distance; (d) the base, tip, and root circles; (e) the line of action; (f) the pressure angle.
6. Define (a) the addendum and dedendum, (b) the pitch, (c) the module.
7. State the basic requirements for a gear material.
8. Explain why a harder material is usually used for the pinion than for the wheel.
9. What materials are used for the manufacture of worm wheels and worms?
10. What types of gear are mitre and crown gears?
11. Describe an involute rack and its uses.
12. Explain the reason for the dedendum being greater than the addendum.
13. What are the advantages of using the involute gear-tooth profile rather than cycloidal profiles?
14. Draw an involute curve generated from an arc subtended by an angle of 30° at the centre of a base circle of 100 mm radius.
15. Draw an approximate involute curve using a pair of compasses on a base circle of 120 mm radius with a module of 20 mm.
16. Calculate the centre distance of a gear pair where the module is 8 mm, the pinion has 20 teeth, and the wheel has 60 teeth.
17. Draw an approximate involute rack with a pinion, using the following data: $m = 20$ mm and $z_1 = 12$.

18. The gear ratio of a gear pair is 3:1, the module is 12 mm, and the pinion has 18 teeth. Determine (a) the pitch-circle diameters, (b) the tip diameters, (c) the root diameters, (d) the circular pitch, (e) the circular tooth thickness, (f) the centre distance of the mating gears.
19. Two parallel shafts are connected by spur gears. The shaft centre distance is about 431 mm. The gear ratio is 4:1. If the pressure angle is 20° and the module is 7 mm, calculate the following: (a) the number of teeth, (b) the pitch-circle diameters, (c) the centre distance, (d) the tip-circle diameters, (e) the root-circle diameters, (f) the angular tooth thickness.
20. The maximum distance at which two gears can operate is given as 225 mm and the gear ratio is 4:3. The module is given as 16 mm and the pressure angle is 20°. Calculate the following: (a) the correct number of teeth to give the gear ratio of 4:3, (b) the pitch-circle diameters, (c) the tip-circle diameters, (d) the root-circle diameters, (e) the base-circle diameters, (f) the working depth of gear teeth, (g) the centre-distance between the gears.
21. Figure 5.18 shows part of the geometry of a spur-gear pair. Name each of the features numbered from 1 to 22.

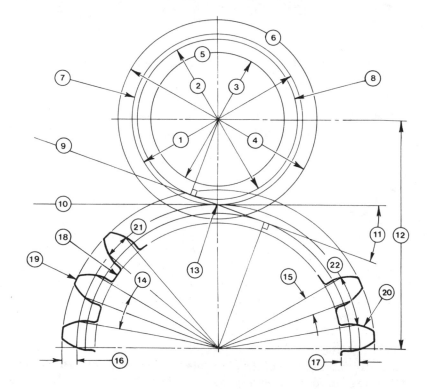

Fig. 5.18

6 Bearings

6.1 Reasons for the use of bearings

A bearing is a support provided to hold a component in its correct position while at the same time allowing it to rotate or slide.

There are two main types of bearing to support different loads (forces):

i) *journal bearings* support radial loads, which act at right angles to the axis of the shaft as shown in fig. 6.1(a);

ii) *thrust bearings* support axial loads, which act along the axis of the shaft as shown in fig. 6.1(b).

Also, bearings may be plain or of the rolling ('antifriction') type.

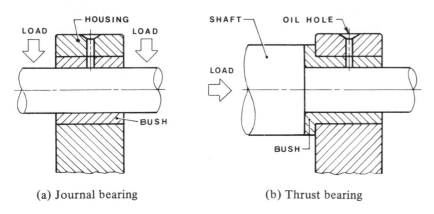

(a) Journal bearing   (b) Thrust bearing

Fig. 6.1 Plain bearings

Friction

Resistance to sliding of two surfaces in contact is called friction and is caused mainly by very small imperfections on the surfaces, even if machined. When magnified many times, these imperfections appear on both surfaces in contact as 'peaks' and 'valleys' and they tend to interlock and resist motion (fig. 6.2(a)). But friction exists even between very smooth surfaces, due to pressure and relative motion of the surfaces.

The effects of friction must be reduced in order to prevent

a) loss of energy in overcoming frictional resistance;

b) overheating of surfaces in contact, which can result in melting and fusing together of mating parts;

c) damage to the surfaces in contact, due to wear.

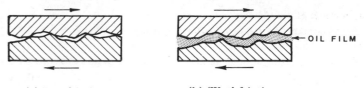

(a) Dry friction   (b) 'Wet' friction

Fig. 6.2 Sliding friction

Lubrication

The effect of dry sliding friction is reduced considerably by introducing a thin film of lubricant – oil or grease – which keeps the two sliding surfaces apart (fig. 6.2(b)).

Oils are liquid in structure and are used mainly for the following:

a) high speeds,

b) high temperatures,

c) as coolant to dissipate frictional heat,

d) in splash and circulation systems of lubrication.

Greases are semisolid in structure and are used mainly for the following:

a) low speeds,

b) corrosion protection of metal bearings,

c) prevention of foreign-matter contamination by partially sealing the bearings,

d) applications where lubricant leakage must be prevented (food, paper, textile industries, etc.),

e) long-life lubrication for sealed or shielded bearings.

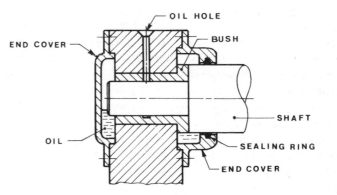

Fig. 6.3 A method of lubricating bearings

6.2 Plain bearings (fig. 6.4)

Plain bearings consist of a supporting part, called the housing, and a mating part which may be a shaft, pivot, or thrust collar.

The ideal plain bearing should be hard, strong, and wear-resistant, with a soft overlay which can be easily deformed, to absorb sudden loads without fracturing, and in which foreign particles can be embedded, thus preventing rapid wear of the mating surfaces.

There are two classes of plain bearings: (a) direct-lined housings and (b) bushes. The materials used for them are discussed in sections 6.3 and 6.4

Direct-lined housings

In this type of plain bearing, the housing is lined directly with bearing material by means of metallurgical bonding or keying. This construction is mostly confined to low-melting-point white metals attached to ferrous housings.

Applications: Cement-mill machinery, crushing plants, large crankshafts, car engines, etc.

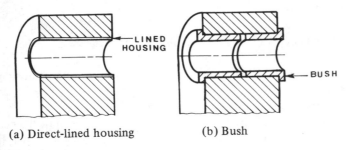

(a) Direct-lined housing   (b) Bush

Fig. 6.4 Plain bearings

Bushes

Bushes are hollow cylindrical pieces which are fitted into a housing to accommodate the mating part. When worn, they are removed and replaced.

To prevent the rotation of bushes in their housings, an interference fit is used. Alternatively, pins, screws, or various shapes of bush ends to fit in correspondingly shaped housing bores are utilised.

There are two types of bush: solid and lined. In both cases the bearing material might be a white metal, a copper alloy, or an aluminium alloy.

Solid bushes   These are made entirely of the bearing material.
Applications: Earth-moving equipment, gearboxes, crankshaft bearings, steering-gear linkages, diesel-engine small-ends, and general applications.

Lined bushes   In these, the bearing material is applied as a lining to a backing material.
Applications: Plant machinery, turbines, large diesel engines, marine gearboxes, etc.

Advantages of plain bearings

a) They usually require only a small radial space.
b) They are cheap to produce.
c) They have vibration-damping properties.
d) They are noiseless in operation.
e) They can be easily machined.
f) They can cope with trapped foreign matter.

Disadvantages of plain bearings

a) They require large supplies of lubricant.
b) They are suitable only for relatively low temperatures and speeds.
c) When starting from rest, the initial resistance to motion is much larger than the running resistance, due to the slow build up of the lubricant film around the bearing surface.

6.3 Bearing alloys

Materials suitable for plain bearings require special properties, as follows:
a) *compressive strength,* to support loads imposed by components;
b) *fatigue strength,* to withstand stresses due to forces which are applied repeatedly and which may vary in magnitude and direction;
c) *corrosion resistance,* to resist corrosion due to oil oxidation products, water, or other contaminants;
d) *ductility,* to be able to deform very slightly in order to absorb sudden loads without fracturing;
e) *thermal conductivity*, to conduct heat generated away from bearing contact surfaces;
f) *low coefficient of friction,* to ensure low friction forces;
g) *embeddability,* so that all trapped foreign particles can sink in below the bearing surface, thus preventing rapid wear of the mating surfaces;
h) *machinability,* for economical and easy bearing production;
j) *bonding property,* to allow the bearing to be attached firmly to a backing material if required.

Bearing alloys have been produced with various combinations of the above properties to suit different applications. The metal on which the alloy is based forms a matrix whose properties are modified by the alloying elements. The harder and stronger the matrix, the higher is its fatigue strength; conversely, the softer it is the better is its embeddability. A soft matrix will accommodate some displacement due to loading and any slight misalignment of a mating part and will allow evenly distributed wear, but it tends to wear more rapidly.

White metals

These alloys, also known as Babbit metals, consist mainly of tin and/or lead with small amounts of antimony and copper. The addition of antimony improves the compressive strength and provides resistance to wear but introduces brittleness. By adding a small amount of copper the desirable ductility and toughness are restored. Lead is used mainly in the interests of cheapness.

White metals have low melting points and are therefore not suitable for high-temperature applications, but their embeddability is very good.

Tin-base alloys are preferable to lead-base alloys as
a) they flow more readily when molten,
b) they shrink less when solidifying,
c) they are more ductile,
d) they are more corrosion-resistant.
However, they are more expensive to produce than lead-base alloys.

A typical composition of a tin-base bearing alloy used for general work would be 86% tin, over 10% antimony, and over 3% copper.

Lead-base alloys are more liable to be corroded by products of oxidation of oil and have a tendency for abrasive wear. They are used for low speeds and light duties.

A typical composition of a lead-base bearing alloy used for low-duty applications would be over 63% lead, 20% tin, 15% antimony, and over 1% copper.

Copper-base alloys
These alloys are harder, stronger, and more resistant to wear than white metals.

Copper—lead alloys Where loads are too high for white metals and sizes of bearings are restricted, an alloy of about 70% copper and 30% lead may be used. About 2% of tin may be added to reduce hardening and brittleness. These types of bearing are used for high speeds, notably in diesel engines.

Lead bronzes These alloys have a bronze matrix consisting of copper and tin, in which the lead is distributed. The hardness and fatigue strength will depend upon the tin content, which may be from 1% to 15%. The lead content may be up to 25%. These bearings have excellent casting properties, are easily machined, and are used for high-duty applications.

Tin bronzes These copper—tin alloys are hard and strong. They have high fatigue and compressive strengths, good resistance to wear, but poor embeddability. They are used for low speeds and high-duty applications.

Phosphor bronzes These alloys have a small phosphorus content which has a deoxidising effect. They are corrosion-resistant, and stronger than simple tin bronzes.

Gun metals These are copper—tin—zinc alloys with occasional small amounts of lead. They are wear- and corrosion-resistant and are used for marine fittings, valves, etc.

Brasses These copper—zinc alloys with very small amounts of aluminium, iron, and manganese are often used as bearings. Bearing brasses have usually about 60% of copper and about 40% of zinc.

Aluminium-base alloys
These alloys may contain up to 20% tin, to provide embeddability, with about 1% of copper and nickel. Their fatigue strength is similar or somewhat higher than that of copper—lead and copper—tin bronzes.

They are mainly used as bearings for diesel engine and motor-car crankshafts and for connecting rods.

Table 6.1 Relative properties of some bearing alloys

Alloy	Hardness	Embedding ability	Corrosion resistance	Fatigue resistance	Thermal conductivity
Tin-base Babbit	Extremely low	Very high	Very high	Low	Medium
Lead-base Babbit	Extremely low	Very high	Medium	Low	Low
Copper—lead alloy	Low	Medium	Low	Medium	Very high
Lead bronze	Low	Medium	Medium	High	Medium
Tin bronze	Low	Low	High	Very high	Medium
Aluminium-base alloy	Very low	Medium	Very high	Very high	High

6.4 Other bearing materials

Cast-iron bearings
A cast-iron bearing is usually simply a hole bored in a cast-iron part to accommodate a mating part. For grey cast-iron, adequate lubrication free from dust is essential.

These types of bearing are in general used for light duties.

Porous metal bearings
Very fine metal powders are partially compressed to a required shape and then sintered at a high temperature. The metal sponges so produced may be filled with a low-friction thermoplastic for dry running or be impregnated with oil.

Porous metals may absorb up to one third of their own volume of lubricant and when running they give off the lubricant, thus wetting the bearing surfaces. When stationary, most of the lubricant is reabsorbed again.

Due to their low compressive strength, these bearings are used for low-duty applications, such as control linkages, door hinges, etc.

Non-metallic bearings

Thermoplastics   Thermoplastics are suitable for injection moulding and extrusion as they can be softened and reshaped with the application of heat. These materials include PTFE (polytetrafluoroethylene), nylon, polystyrene, polypropylene, etc.

Applications include bushes for medical equipment, food-preparation equipment, textile machinery, pumps, etc. PTFE, with its high thermal resistance and load-carrying capacity, is used for chemical pumps, railway-point pivot bushes, conveyor bushes, etc.

Thermosetting plastics   Thermosets are chemically changed when heated — they become rigid, and this change cannot be reversed. For use as bearings these materials are always reinforced by the addition of asbestos, silicone resins, carbon, or metals.

These types of bearing are used for hot-duct supports, jet-pipe supports, cranes, vibratory rollers, pump bushes, etc.

Carbon (graphite)   Graphite bearings have a very high resistance to elevated temperatures, need no lubrication, have a polishing effect on the mating parts, and can run in fluids which attack other bearing materials.

Carbon bearings are used for food and textile machinery, furnace and boiler equipment, chemical agitators, bottle-washing plant, trolley wheels, etc.

Rubber   Rubber bearings are easily deflected, thus reducing stresses and damping vibrations. They are usually used for pumping purposes with water as the lubricant. Rusting mating surfaces should be avoided, as rough surfaces will damage the rubber.

Wooden bearings   Lignum vitae is one of the hardest and densest of all woods. It is very strong in compression, can resist the action of certain chemicals, and is used for bearings in food- and chemical-processing machinery, etc.

Hard maple when impregnated with oil is used for textile machinery, loose pulleys, etc.

Jewel bearings are used for instruments, clock mechanisms, etc.

6.5 Ball and roller rolling bearings

In ball and roller bearings, the sliding friction of plain bearings is replaced by a much lower rolling friction.

Also, when starting plain bearings from rest, their initial resistance to motion is considerably larger than the running resistance after the lubricant film has been built up around the bearing surface. However, when starting ball and roller bearings from rest, their initial resistance to motion is only slightly more than their resistance to continued running. These bearings are therefore used for devices which are subject to frequent starting and stopping.

Bearing parts (fig. 6.5)
Ball and roller bearings consist of
a) an *inner ring,* which fits on the shaft;
b) an *outer ring,* which fits inside the housing;
c) *balls or rollers,* which provide a rolling action between the rings;
d) a *cage,* which separates adjacent ball surfaces, which rotate in opposite relative directions, and prevents sliding friction between them.

Note: Rings sometimes are called 'races'.

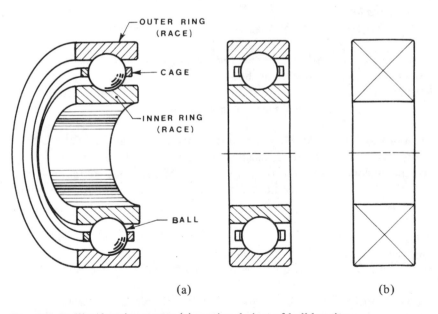

Fig. 6.5 Rolling-bearing parts: (a) sectional view of ball bearing, (b) conventional representation of a bearing

Bearing materials

The materials used for rolling bearings are high-carbon chromium steels which are very hard and resistant to wear. Surfaces in contact must be highly polished to reduce wear and to provide smooth rolling movement without any sliding.

The cages are made of low-carbon steels, bronzes, or brasses, though for high-temperature applications case-hardened and stainless steels are used.

Fits

The inner ring of a journal bearing must have an interference fit on a revolving shaft, to prevent creep. (Creep is the slow rotary movement of the inner ring relative to the rotating shaft or, alternatively, the slow rotary movement of the outer ring relative to the rotating housing.)

The inner ring is usually held axially between a shaft shoulder and a nut, as shown in fig. 6.6(a) and (b).

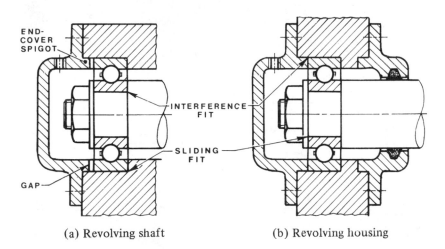

(a) Revolving shaft   (b) Revolving housing

Fig. 6.6 Assembly fits for rolling bearings

The outer ring is assembled inside the stationary housing, with a sliding fit to permit correct axial position without preload. Often the outer ring is not held axially — a clearance is introduced in order to allow for inaccurate machining and for relative expansion or contraction of the shaft and the housing due to temperature changes, as shown in fig. 6.6(a).

For revolving parts on the shaft, such as wheels, pulleys, etc., the outer rings have an interference fit inside the housing of these wheels, whereas the inner rings have a sliding fit on the stationary shaft, as shown in fig. 6.6(b).

General rule: Rotating rings require interference fits; stationary rings require sliding fits.

6.6 Types of rolling bearing

Single-row deep-groove ball bearings (fig. 6.7(a))

These bearings incorporate a deep hardened raceway or track which makes them suitable for radial and axial loads in either direction, providing the radial loads are greater than the axial loads.

These bearings are self-contained units — they can be handled and assembled as a single component. Also, they may be of the prelubricated types, having integral seals or shields which retain the grease within the bearing and prevent foreign matter entering it (fig. 6.11).

Applications These bearings can be used as locating bearings for high-speed spindles, motorcycle engines, electric motors, circular saws, turbine shafts, wood cutters, gearboxes, pulleys, pumps, etc.

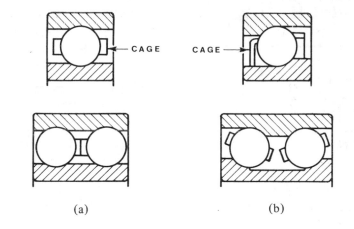

(a)   (b)

Fig. 6.7 (a) Single- and double-row deep-groove ball bearings,
(b) single- and double-row angular-contact ball bearings

Single-row angular-contact ball bearing (fig. 6.7(b))

Outer rings and sometimes inner rings are machined with high and low shoulders to take one-directional thrust or combined radial and axial loading. To support thrust loads in either direction, these bearings can be mounted in opposing pairs, suitably adjusted to prevent end play or preloading due to over-adjustment.

Double-row angular-contact ball bearings can be used as alternatives to opposing pairs of single-row bearings, simplifying the mounting and saving space.

Applications These bearings are used for high-speed spindles for boring, milling, and drilling machines and for motorcycle engines, worm gears, hot-gas fans, radar aerials, etc.

Double-row self-aligning ball bearings (fig. 6.8(a))

These bearings are designed for the cases where alignment of inner and outer rings cannot be assured during assembly or in service. The inner ring has two deep-groove raceways, whereas the outer ring has a single spherical raceway. This allows the inner and outer rings to be misaligned relative to each other.

Applications These bearings should not be used for very heavy radial loads and they will not support any axial loads. Protective shields cannot be fitted. These bearings are used for gearboxes, air blowers, cutter shafts of planing machines, etc.

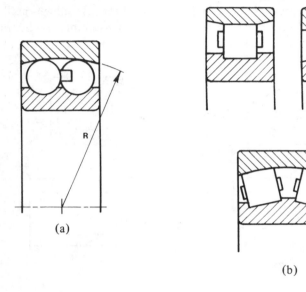

(a)

(b)

Fig. 6.8 (a) Double-row self-aligning ball bearing, (b) single- and double-row roller bearings

Single-row roller bearings (fig. 6.8(b))

Roller bearings are usually detachable units. They have a greater load-carrying capacity than ball bearings of equivalent size as they make line contact rather than point contact with their rings. These bearings are not suitable for axial loading. A slight axial displacement is permissible between the inner and outer rings. These types of bearing are cheaper to manufacture than equivalent ball bearings.

Applications These bearings are suitable for heavy and sudden loading, high speeds, and continuous service. They are used for vibrating motors, stone crushers, dredging machinery, ship propellers and rudder shafts, belt conveyors, locomotive axles, flywheels, crankshafts, presses, etc.

Tapered-roller bearings (fig. 6.9(a))

These bearings are so designed that, when projected, construction lines corresponding to the surfaces of contact of rollers and rings will meet at a *apex* point on the bearing axis.

Like angular-contact ball bearings, these bearings will carry a combination of radial and single-direction thrust loads. The inner ring, called the cone, with a tapered roller and cage, is assembled as a complete unit; whereas the outer ring is detachable.

Two bearings can be mounted on a shaft, but they must be accurately adjusted axially to ensure the proper running clearance between the roller and the outer ring, called the cup.

Applications These bearings are suitable for lathe spindles, bevel-gear transmissions, gearboxes for heavy trucks, car drives, car wheels, etc.

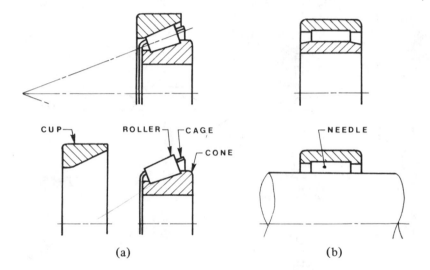

(a)

(b)

Fig. 6.9 (a) Tapered roller bearing (b) Needle roller bearings

Needle-roller bearings (fig. 6.9(b))

These bearings are fitted with small-diameter rollers and are used for radial loads at slow speeds and oscillating motion. They are especially suitable for restricted spaces and where the weight of components is critical, as in the aircraft industry.

Sometimes, due to space limitations, the inner ring is not used; instead, the shaft is hardened and ground and is used directly as an inner ring.

Applications These bearings are used for aircraft applications, live tailstock centres, bench-drill spindles, light gearboxes, etc.

Thrust ball bearings (fig. 6.10(a))

These bearings can only take thrust loads. They consist of two loose thrust rings grooved to accommodate the balls with their corresponding cage. One of the rings has a smaller bore than the other and engages on the shaft, while the other ring has a larger outer diameter which engages in the housing. A *double-thrust bearing* has three rings.

Applications Thrust ball bearings are used for heavy axial loads and low speeds, and are suitable for thrust spindles, tailstock centres for heavy work, vertical shafts, pillars supporting heavy vertical loads, etc.

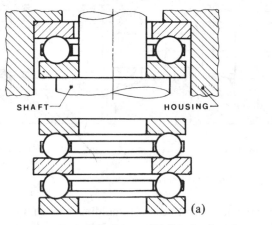

SHAFT HOUSING

(a) (b)

Fig. 6.10 (a) Thrust ball bearings (b) Duplex bearing

Duplex bearings (fig. 6.10(b))

Duplex bearings are used to support thrust loads in either direction and have a split outer or inner ring. They should be used for thrust loads alone or for combined thrust and radial loads only when the thrust load is very much greater than the radial one.

Prelubricated bearings (fig. 6.11)

These bearings incorporate metal shields and/or seals, which are usually fastened to the outer ring. The close clearance between the seal and the inner ring retains the long-life grease within the bearing and prevents foreign matter entering from outside.

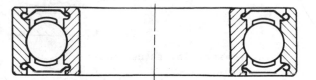

Fig. 6.11 Prelubricated sealed ball bearing

Table 6.2 Comparison of different types of rolling bearing

	Type of bearing	Approximate coefficient of friction	Radial-load capacity	Axial-load capacity
	Single-row deep-groove ball bearing (Double-row)	0.001	Light and medium (Heavy)	Light and medium (Medium)
	Single-row angular-contact ball bearing	0.002	Medium	Medium
	Double-row angular-contact ball bearing	0.002	Medium and heavy	Medium
	Double-row self-aligning ball bearing	0.001	Medium	Light
	Single-row roller bearing (Double-row)	0.001	Heavy (Very heavy)	–
	Tapered-roller bearing	0.002	Heavy	Medium and heavy
	Needle-roller bearing	0.003	Heavy	–
	Single-row thrust ball bearing	0.001	–	Light and medium
	Double-row thrust ball bearing	0.001	–	Heavy and medium
	Duplex ball bearing	0.002	Very light	Medium

6.7 Design of rolling-bearing assemblies

Ball and roller bearings are normally mounted on a shaft, with the inner ring clamped against the shaft shoulder by means of a lock nut.

The outer ring is sometimes fixed endways between the spigot of the end cover and the housing shoulder, as shown in fig. 6.12.

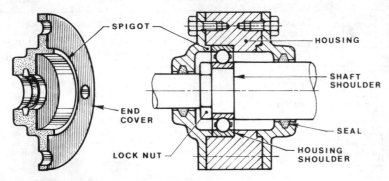

Fig. 6.12 Ball-bearing assembly

To ensure that a bearing operates properly it must be protected against foreign matter entering the housing, and at the same time the lubricant must be kept inside the bearing. This function is performed by the bearing seals or shields (fig. 6.11), or by providing the end covers with contact or non-contact seals, as shown in figs 6.12 and 6.13.

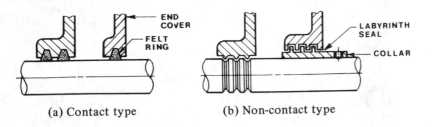

(a) Contact type   (b) Non-contact type

Fig. 6.13 Bearing seals

The contact type of seal usually consists of a ring made of felt, leather, synthetic rubber, etc., fixed into the groove in the end cover and contacting the rotating shaft.

The non-contact type of seal is used when high temperatures and speeds are employed. A close clearance is provided by a series of grooves in the end cover and in the shaft or shaft-collar which are then filled with sealing grease. This method eliminates the friction and wear of the rubbing contact-type seals.

When a shaft is supported by two bearings, only one of the bearings should be fixed axially. The other bearing must provide axial adjusting movement by the ring sliding to accommodate the tolerances of position and to allow for relative differences in axial dimensions between the shaft and the housing due to temperature changes, as shown in fig. 6.14.

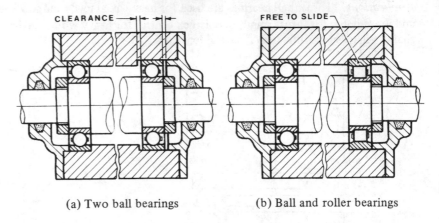

(a) Two ball bearings   (b) Ball and roller bearings

Fig. 6.14 Shaft supported by two bearings

In fig. 6.15, the thrust load due to the workholding is taken up by a thrust ball bearing. The feed-motion thrust is carried by a tapered-roller bearing, which also supports the cutting loads. Due to space limitation, a needle bearing is used which also provides the axial adjusting movement.

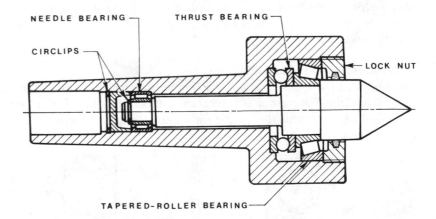

Fig. 6.15 Live tailstock centre of a lathe

6.8 Test questions

1. Explain briefly the reasons for the use of bearings and say why the effects of friction on bearings must be reduced.
2. Discuss briefly the beneficial effect of using lubricants in bearings and state the relative advantages of using oil or grease.
3. What are the basic differences between journal and thrust bearings and between plain and rolling bearings?
4. Explain briefly the difference between direct-lined housings and solid and lined bushes.
5. Discuss the advantages and disadvantages of plain bearings.
6. List the special properties required for materials suitable for plain bearings.
7. Discuss briefly the following alloys and state their relative merits as bearing materials: (a) tin-base and lead-base Babbit metals, (b) copper–lead alloys, (c) tin bronzes, (e) aluminium-base alloys.
8. Discuss briefly the following bearings and suggest their industrial applications: (a) cast-iron bearings, (b) porous metal bearings, (c) PTFE and nylon bearings, (d) graphite bearings, (e) rubber bearings.
9. State the main advantages of using rolling bearings.
10. Sketch and name the four important parts of a roller bearing, and sketch a conventional representation of any bearing.
11. What materials are used for the manufacture of rolling-bearing parts?
12. Figure 6.16 shows different types of rolling bearing. Identify each of these bearings, stating the type of loading each will support and indicating suitable industrial applications.

13. State the types of fit employed when assembling rolling bearings for revolving shafts and stationary housings, and for stationary shafts and revolving housings. Give reasons for using these fits.
14. Name the rolling bearings which are *not* suitable to support the following: (a) thrust (axial) loads, (b) journal (radial) loads.
15. Name the rolling bearings which are suitable to support the following:
    a) a combination of thrust and journal loads in one direction only,
    b) a combination of thrust and journal loads in both directions.
16. Name the types of bearing which would be used when (a) the radial bearing space is limited, (b) the shaft alignment cannot be guaranteed, (c) the bearing cannot be lubricated regularly.
17. Name two types of lubricant seal, and indicate one that is suitable for high temperature and speed applications.
18. Explain why only one of the two separate ball bearings supporting a shaft should be fixed axially. What type of bearing is suitable to replace one of these ball bearings?
19. Complete the sub-assembly drawing shown in fig. 6.17 by locating axially the inner and outer rings and incorporating the lubricant seals. Tracing paper may be used.

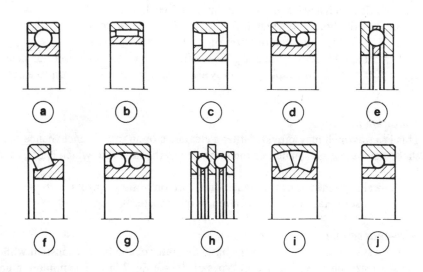

Fig. 6.16 Different types of rolling bearing

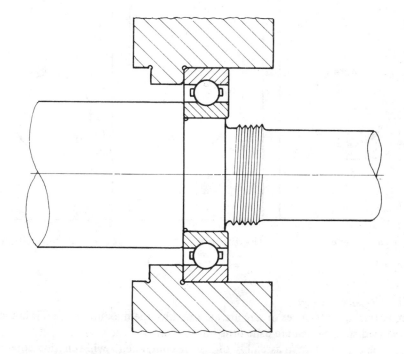

Fig. 6.17 Sub-assembly drawing of a rolling-bearing mounting

7 Tolerancing

It is impossible to manufacture a component to an exact design size or shape. To take account of this, a tolerance is permitted or 'tolerated'.

The tolerance is intended to allow for reasonable inaccuracy in manufacturing or positioning and is defined as the maximum amount of deviation from the given basic design size (or margin of error) which is permissible if the component or assembly is to function as planned.

Figure 7.1 shows three possible errors when producing a hole in a component:
a) Error of size — the hole may be larger or smaller than required.
b) Error of form — the required circular hole may be non-circular. This is a geometrical error.
c) Error of position — the centre of the hole may not be in the required position. This is also a geometrical error.

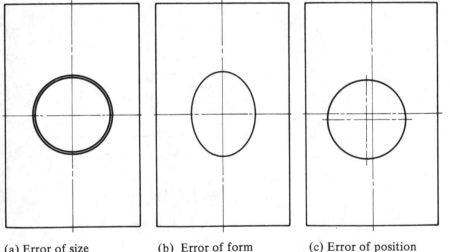

(a) Error of size   (b) Error of form   (c) Error of position

Fig. 7.1

7.1 Tolerances of size

On a drawing, tolerances of size are indicated by the maximum and minimum permitted sizes, which are called the limits of size.

To be certain that an assembly of mating components will function correctly, a designer must ensure that all parts will fit together in the required manner.

A particular fit will depend solely on the prescribed maximum and minimum limits of size of the two separate components which are to be assembled.

Engineering fits can be divided into three main types: clearance fits, interference fits, and transition fits.

Clearance fit

This is a fit which provides a clearance; hence the shaft is always smaller than the hole into which it fits, as in fig. 7.2(a). *Clearance* is the positive difference between the sizes of the hole and the shaft.

Typical applications of the clearance fit are on rotating shafts, loose pulleys, fast pulleys, bearings, cross-head slides, etc.

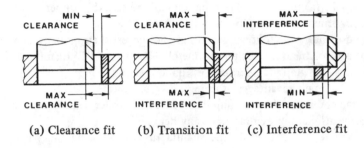

(a) Clearance fit   (b) Transition fit   (c) Interference fit

Fig. 7.2 Engineering fits

Interference fit

This is a fit which always provides an interference; hence the shaft is always bigger than the hole into which it fits, as in fig. 7.2(c). *Interference* is the negative difference between the sizes of the hole and the shaft.

Typical applications of the interference fit are on pressed-in bushes or sleeves, crank pins, shrunk-on couplings, iron tyres, railway wheels shrunk on to axles, etc.

Transition fit

This is a fit which may provide either a clearance or an interference; hence the shaft may be bigger, smaller, or the same size as the hole into which it fits, as in fig. 7.2(b).

Typical applications of the transition fit are on bushes, spigots, fasteners, pins, keys, stationary parts for location purposes, etc.

Hole-basis system

This is a system of fits in which the basic diameter of the hole is constant while the shaft size varies with different types of fit, see fig. 7.3(a). The minimum limit of hole size is the basic size.

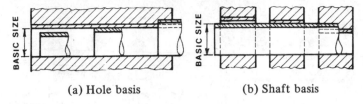

(a) Hole basis   (b) Shaft basis

Fig. 7.3 Systems of fits

The hole basis is more economical than the shaft basis as only one size of drill or reamer need be used to produce different fits, the shafts being turned and ground to the required sizes, thus making manufacture and measurement much easier.

Shaft-basis system

This is a system of fits in which the hole size is varied to produce the required type of fit, with the basic diameter of the shaft being constant. The maximum limit of the shaft is the basic size, see fig. 7.3(b).

This system tends to be less economical, as a series of drills is required. It is usually adopted where a single driving shaft accommodates a number of pulleys, bearings, collars, couplings, etc.

British Standard BS 4500, *ISO limits and fits,* gives a selection of hole and shaft tolerances to cover a wide range of engineering applications.

For a selected range of fits which is adequate for most practical requirements, the BS 4500A and BS 4500B data sheets give the fits on a hole and shaft basis respectively.

For most general applications the hole-basis fits are usually recommended. Data sheet BS 4500A, fig. 7.9, shows a range of fits derived from selected hole tolerances (H11, H9, H8, H7) and shaft tolerances (c11, d10, e9, f7, g6, h6, k6, n6, p6, s6), where capital letters refer to holes, lower-case letters to shafts, and greater numbers to bigger tolerances.

Determining working limits

We will use the data sheet BS 4500A, fig. 7.9, to determine the working limits for the assembly shown in fig. 7.4(a), assuming this to be a designer's layout which has been passed to a detail draughtsman for preparation of the working drawings of the components.

In order to decide on a desirable fit, we must consider the function of the assembly. The shaft is going to rotate in the bush; hence a clearance fit is required, allowing sufficient space for lubricant, but not so much as to cause wobbling of the shaft. The most suitable fit will be H8/f7, as shown in fig. 7.4(b).

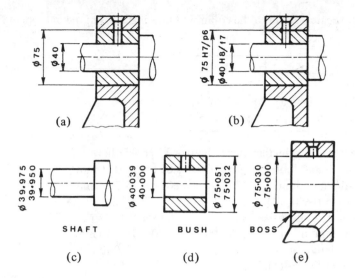

(a)   (b)

SHAFT   BUSH   BOSS

(c)   (d)   (e)

Fig. 7.4 Method of size tolerancing

We now locate 40 mm in the column of nominal sizes in the data sheet BS 4500A in fig. 7.9, remembering that column 'Over' means 'over but excluding' and 'To' means 'to and including'.

The required tolerances are shown below:

Over	To	H8	f7
30	40	+39	−25
		0	−50

As all tolerances are given in micrometres (0.000 001 m or 0.001 mm), the shaft basic size 40 f7 will have

    maximum limit   40.000 − 0.025 = 39.975 mm

and   minimum limit   40.000 − 0.050 = 39.950 mm

as shown in fig. 7.4(c).

(When tolerancing, the same number of decimal places must be used for both limits.)

The bush hole basic size 40 H8 will have

    maximum limit   40.000 + 0.039 = 40.039 mm

and   minimum limit   40.000 + 0   = 40.000 mm

as shown in fig. 7.4(d).

Assuming that the bush is going to be pressed into the bracket boss, then an interference fit will be appropriate. The H7/p6 fit seems to be most suitable, as shown in fig. 7.4(b).

Now locate 75 mm in the column of nominal sizes in the data sheet BS 4500A, fig. 7.9:

Over	To	H7	p6
65	80	+30	+51
		0	+32

The bush outside-diameter basic size 75 p6 will have
  maximum limit   75.000 + 0.051 = 75.051 mm
and   minimum limit   75.000 + 0.032 = 75.032 mm
as shown in fig. 7.4(d).
  The bracket boss basic size 75 H7 will have
  maximum limit   75.000 + 0.030 = 75.030 mm
and   minimum limit   75.000 + 0   = 75.000 mm
as shown in fig. 7.4(e).

Dimensioning tolerances

A toleranced drawing of a rectangular component is shown in fig. 7.5(a) and how it is interpreted is shown in fig. 7.5(b). The tolerance zones of 0.6 mm shown are the differences between the maximum and minimum limits.

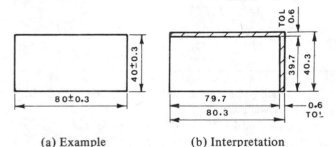

(a) Example   (b) Interpretation

Fig. 7.5 Tolerancing a component

Tolerance limits between centres of holes can be indicated either by *chain dimensioning,* as in fig. 7.6(a), or by *progressive dimensioning* from a common datum as shown in fig. 7.7(a).
  The use of chain dimensions results in an accumulation of tolerances between the holes and the edge of the plate, and this may endanger the functional requirements.

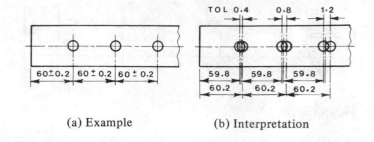

(a) Example   (b) Interpretation

Fig. 7.6 Chain dimensioning

Progressive dimensioning from a common datum on the component prevents this accumulation of tolerances, as each hole is toleranced directly from the datum.

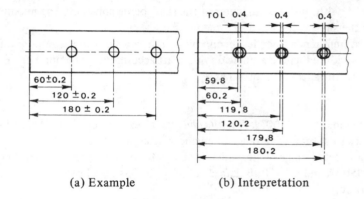

(a) Example   (b) Intepretation

Fig. 7.7 Progressive dimensioning

When tolerancing an individual linear dimension, the method of specifying directly maximum and minimum limits of size is preferable. The larger limit should be given first, and the same number of decimal places should be indicated for both limits.
  Four correct methods of tolerancing are shown in fig. 7.8.

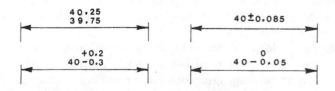

Fig. 7.8 Methods of tolerancing linear dimensions

44

SELECTED ISO FITS—HOLE BASIS

Diagram to scale for 25 mm diameter — Clearance fits: H11/c11, H9/d10, H9/e9, H8/f7, H7/g6, H7/h6. Transition fits: H7/k6, H7/n6. Interference fits: H7/p6, H7/s6. Holes / Shafts.

Nominal sizes		Clearance fits												Transition fits				Interference fits				Nominal sizes	
Over	To	H11	c11	H9	d10	H9	e9	H8	f7	H7	g6	H7	h6	H7	k6	H7	n6	H7	p6	H7	s6	Over	To
mm	mm	0·001 mm	0·001 mm	0·001 mm	0·001 mm	0·001 mm	0·001 mm	0·001 mm	0·001 mm	0·001 mm	0·001 mm	0·001 mm	0·001 mm	0·001 mm	0·001 mm	0·001 mm	0·001 mm	0·001 mm	0·001 mm	0·001 mm	0·001 mm	mm	mm
—	3	+60 / 0	−60 / −120	+25 / 0	−20 / −60	+25 / 0	−14 / −39	+14 / 0	−6 / −16	+10 / 0	−2 / −8	+10 / 0	−6 / 0	+10 / 0	+6 / +0	+10 / 0	+10 / +4	+10 / 0	+12 / +6	+10 / 0	+20 / +14	—	3
3	6	+75 / 0	−70 / −145	+30 / 0	−30 / −78	+30 / 0	−20 / −50	+18 / 0	−10 / −22	+12 / 0	−4 / −12	+12 / 0	−8 / 0	+12 / 0	+9 / +1	+12 / 0	+16 / +8	+12 / 0	+20 / +12	+12 / 0	+27 / +19	3	6
6	10	+90 / 0	−80 / −170	+36 / 0	−40 / −98	+36 / 0	−25 / −61	+22 / 0	−13 / −28	+15 / 0	−5 / −14	+15 / 0	−9 / 0	+15 / 0	+10 / +1	+15 / 0	+19 / +10	+15 / 0	+24 / +15	+15 / 0	+32 / +23	6	10
10	18	+110 / 0	−95 / −205	+43 / 0	−50 / −120	+43 / 0	−32 / −75	+27 / 0	−16 / −34	+18 / 0	−6 / −17	+18 / 0	−11 / 0	+18 / 0	+12 / +1	+18 / 0	+23 / +12	+18 / 0	+29 / +18	+18 / 0	+39 / +28	10	18
18	30	+130 / 0	−110 / −240	+52 / 0	−65 / −149	+52 / 0	−40 / −92	+33 / 0	−20 / −41	+21 / 0	−7 / −20	+21 / 0	−13 / 0	+21 / 0	+15 / +2	+21 / 0	+28 / +15	+21 / 0	+35 / +22	+21 / 0	+48 / +35	18	30
30	40	+160 / 0	−120 / −280	+62 / 0	−80 / −180	+62 / 0	−50 / −112	+39 / 0	−25 / −50	+25 / 0	−9 / −25	+25 / 0	−16 / 0	+25 / 0	+18 / +2	+25 / 0	+33 / +17	+25 / 0	+42 / +26	+25 / 0	+59 / +43	30	40
40	50	+160 / 0	−130 / −290																			40	50
50	65	+190 / 0	−140 / −330	+74 / 0	−100 / −220	+74 / 0	−60 / −134	+46 / 0	−30 / −60	+30 / 0	−10 / −29	+30 / 0	−19 / 0	+30 / 0	+21 / +2	+30 / 0	+39 / +20	+30 / 0	+51 / +32	+30 / 0	+72 / +53	50	65
65	80	+190 / 0	−150 / −340																	+30 / 0	+78 / +59	65	80
80	100	+220 / 0	−170 / −390	+87 / 0	−120 / −260	+87 / 0	−72 / −159	+54 / 0	−36 / −71	+35 / 0	−12 / −34	+35 / 0	−22 / 0	+35 / 0	+25 / +3	+35 / 0	+45 / +23	+35 / 0	+59 / +37	+35 / 0	+93 / +71	80	100
100	120	+220 / 0	−180 / −400																	+35 / 0	+101 / +79	100	120
120	140	+250 / 0	−200 / −450	+100 / 0	−145 / −305	+100 / 0	−84 / −185	+63 / 0	−43 / −83	+40 / 0	−14 / −39	+40 / 0	−25 / 0	+40 / 0	+28 / +3	+40 / 0	+52 / +27	+40 / 0	+68 / +43	+40 / 0	+117 / +92	120	140
140	160	+250 / 0	−210 / −460																	+40 / 0	+125 / +100	140	160
160	180	+250 / 0	−230 / −480																	+40 / 0	+133 / +108	160	180
180	200	+290 / 0	−240 / −530	+115 / 0	−170 / −355	+115 / 0	−100 / −215	+72 / 0	−50 / −96	+46 / 0	−15 / −44	+46 / 0	−29 / 0	+46 / 0	+33 / +4	+46 / 0	+60 / +31	+46 / 0	+79 / +50	+46 / 0	+151 / +122	180	200
200	225	+290 / 0	−260 / −550																	+46 / 0	+159 / +130	200	225
225	250	+290 / 0	−280 / −570																	+46 / 0	+169 / +140	225	250
250	280	+320 / 0	−300 / −620	+130 / 0	−190 / −400	+130 / 0	−110 / −240	+81 / 0	−56 / −108	+52 / 0	−17 / −49	+52 / 0	−32 / 0	+52 / 0	+36 / +4	+52 / 0	+66 / +34	+52 / 0	+88 / +56	+52 / 0	+190 / +158	250	280
280	315	+320 / 0	−330 / −650																	+52 / 0	+202 / +170	280	315
315	355	+360 / 0	−360 / −720	+140 / 0	−210 / −440	+140 / 0	−125 / −265	+89 / 0	−62 / −119	+57 / 0	−18 / −54	+57 / 0	−36 / 0	+57 / 0	+40 / +4	+57 / 0	+73 / +37	+57 / 0	+98 / +62	+57 / 0	+226 / +190	315	355
355	400	+360 / 0	−400 / −760																	+57 / 0	+244 / +208	355	400
400	450	+400 / 0	−440 / −840	+155 / 0	−230 / −480	+155 / 0	−135 / −290	+97 / 0	−68 / −131	+63 / 0	−20 / −60	+63 / 0	−40 / 0	+63 / 0	+45 / +5	+63 / 0	+80 / +40	+63 / 0	+108 / +68	+63 / 0	+272 / +232	400	450
450	500	+400 / 0	−480 / −880																	+63 / 0	+292 / +252	450	500

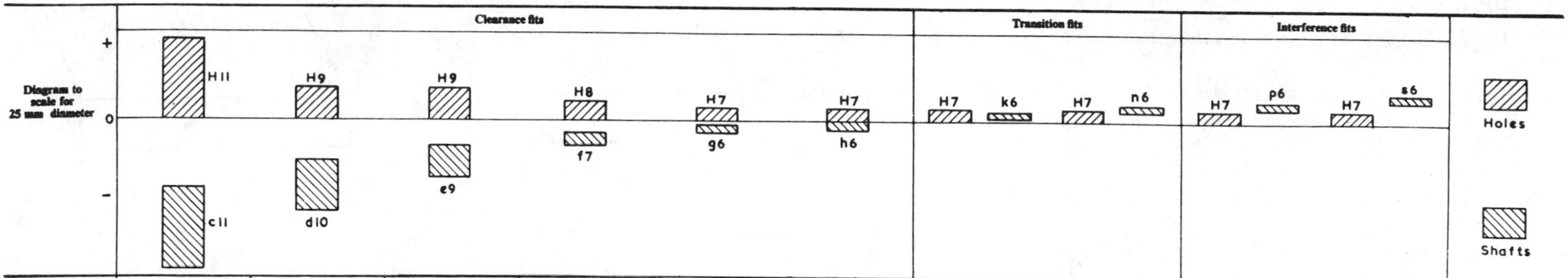

Fig. 7.9 British Standard data sheet BS 4500A: selected ISO fits — hole basis

7.2 Geometrical tolerances

In certain circumstances, tolerances of size are not always sufficient to provide the required control of form:

a) in fig. 7.10(a) the shaft has the same diameter measurement in all possible positions but is not circular;

b) in fig. 7.10(b) the component has the same thickness throughout but is not flat;

c) in fig. 7.10(c) the component is circular in all cross-sections but is not straight.

The form of these components can be controlled by means of geometrical tolerances.

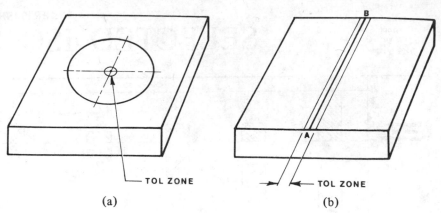

(a)   (b)

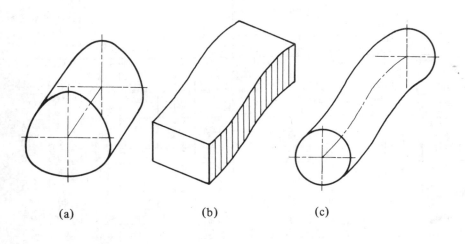

(a)   (b)   (c)

Fig. 7.10 Errors of form

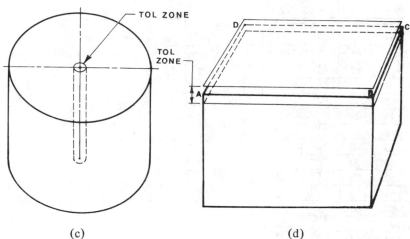

(c)   (d)

Fig. 7.11 Tolerance zones

Geometrical tolerance is defined as the maximum permissible overall variation of *form* or *position* of a feature.

Geometrical tolerances are used

i) to specify the required accuracy in controlling the form of a feature,

ii) to ensure correct functional positioning of a feature,

iii) to ensure the interchangeability of components, and

iv) to facilitate the assembly of mating components.

The *tolerance zone* is an imaginary area or volume within which the controlled feature of the manufactured component must be completely contained.

Figure 7.11 shows four components which, after being inspected, will be passed as correctly manufactured if the features controlled lie within the given tolerance zones.

a) In fig. 7.11(a), the centre of the circle is required to lie within the tolerance area indicated.

b) In fig. 7.11(b), the scribed line AB is required to lie within the tolerance area between two parallel lines.

c) In fig. 7.11(c), the axis of the cylinder is required to lie within the tolerance cylinder indicated.

d) In fig. 7.11(d), the surface ABCD is required to lie within the tolerance volume.

7.3 Indicating geometrical tolerances on drawings

To eliminate the need for descriptive notes, geometrical tolerances are indicated on drawings by symbols, tolerances, and datums, all contained in compartments of a rectangular frame, as shown in fig. 7.12.

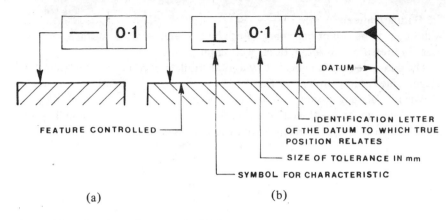

(a)   (b)

Fig. 7.12 Indication of geometrical tolerances

Type of tolerance and tolerance symbols
The main types of geometrical tolerance are of form, attitude, and location.

Form tolerance   This specifies the required geometric shape of a single feature.

Table 7.1   Tolerances of form

Characteristic to be toleranced	Symbol	Abbreviation
Straightness	—	STR TOL
Flatness	▱	FLAT TOL
Roundness	○	RD TOL
Cylindricity	⌀	CYL TOL
Profile of a line	⌒	—
Profile of a surface	⌓	—

Attitude tolerance   This specifies the required orientation of a feature relative to a datum.

Table 7.2   Tolerances of attitude

Characteristic to be toleranced	Symbol	Abbreviation
Parallelism	//	PAR TOL
Squareness	⊥	SQ TOL
Angularity	∠	ANG TOL

Location tolerance   This specifies the required position of a feature relative to a datum.

Table 7.3   Tolerances of location

Characteristic to be toleranced	Symbol	Abbreviation
Position	⊕	POSN TOL
Concentricity	◎	CONC TOL
Symmetry	═	SYM TOL

7.4 Advantages of using geometrical tolerances
1. Geometrical tolerances convey very briefly and precisely the complete geometrical requirements on engineering drawings.
2. The use of symbols and boxes eliminates the need for lengthy descriptive notes and corresponding dimensions; therefore the drawings are much clearer to read.
3. The symbols used are internationally recommended; hence the language barrier is minimised and misunderstanding is eliminated.
4. One type of geometrical tolerance can control another form. For instance, squareness can control flatness and straightness.

7.5 Feature controlled
The feature controlled by a geometrical tolerance is indicated by an arrowhead at the end of a leader line from the tolerance frame. When the tolerance refers to an outline of a feature or to a surface represented by an outline, the arrowhead may touch either the outline, as shown in fig. 7.13(a), or an extension line from the outline, as shown in fig. 7.13(b), but *not* at a dimension line as shown in fig. 7.13(c).

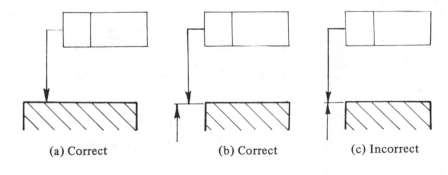

(a) Correct   (b) Correct   (c) Incorrect

Fig. 7.13 Indication of feature controlled (outline or surface only)

47

If the tolerance refers to the axis or median (central) plane of a single-feature part, the arrowhead may touch either the axis or median plane itself, fig. 7.14(a), or the dimension line relevant to the feature whose axis is concerned, fig. 7.14(b).

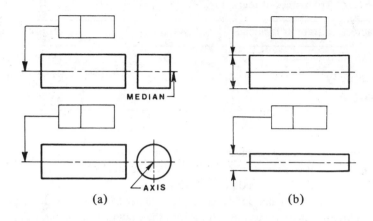
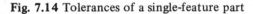

Fig. 7.14 Tolerances of a single-feature part

If the tolerance refers to the axis or median plane of only one feature of a multi-feature part, the arrowhead touches the dimension line relevant to the feature whose axis is concerned, as shown in fig. 7.15(a).

If the tolerance refers to the common axis or median plane of a number of features, the arrowhead touches the axis or median plane, as shown in fig. 7.15(b).

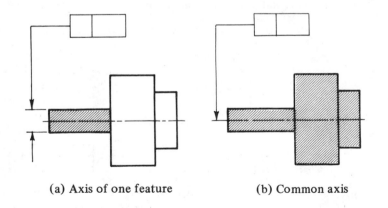

(a) Axis of one feature   (b) Common axis

Fig. 7.15 Tolerances of multi-feature part

7.6 Datum features

A datum may be real or imaginary. It can be a reference plane, surface, or axis used for measuring, location, or inspection purposes.

The datum feature is indicated by a solid equilateral triangle at the end of a leader line from the tolerance frame.

When the datum feature is an outline or a surface represented by an outline, the triangle may be positioned either on the outline or on an extension line from the outline (but not at a dimension line), as shown in fig. 7.16.

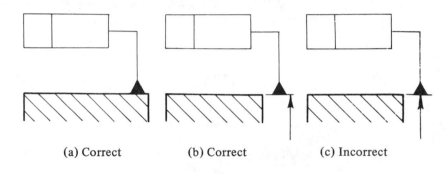

(a) Correct   (b) Correct   (c) Incorrect

Fig. 7.16 Indication of datum feature (outline or surface only)

If the datum feature is the axis or median plane of a single-feature part, the triangle may be positioned either on the axis or median plane itself or on a projection line at the dimension line relevant to the feature whose axis is concerned, as shown in fig. 7.17(a).

If the datum feature is the axis or median plane of a particular feature of a multi-feature part, the triangle is positioned on a projection line at the dimension line relevant to the feature whose axis is concerned, as shown in fig. 7.17(b).

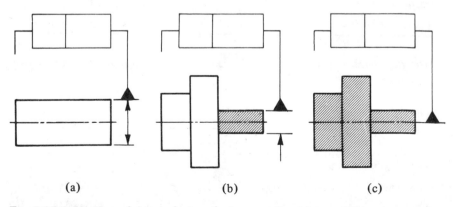

(a)   (b)   (c)

Fig. 7.17 Indication of datum feature (axis or median plane only)

If the datum feature is the common axis or median plane of a number of features, the triangle is positioned on the axis or median plane as shown in fig. 7.17(c).

If the datum feature cannot be clearly and simply connected to the tolerance frame, then a capital letter is connected to the datum feature and is referred to in a separate part of the tolerance frame as shown in fig. 7.18.

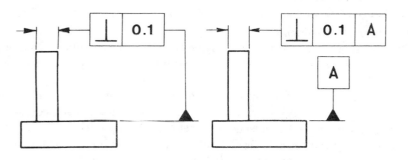

Fig. 7.18 Datum reference

7.7 Boxed dimensions

Dimensions enclosed in a box, e.g. 40 or ⌀ 20, define the true position or exact location of a feature on a component.

7.8 General principles of geometrical tolerancing

1. Geometrical tolerances apply to the whole length or surface of the feature unless stated or indicated otherwise. Figure 7.19 shows how a geometrical tolerance is limited to a particular part of the feature.
2. The use of geometrical tolerances does not imply the use of any particular method of production or inspection.
3. A line or surface of a feature controlled by geometrical tolerances may be of any form and may take any position provided it remains within its tolerance zone.

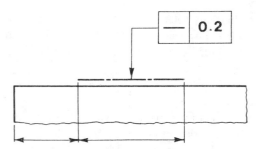

Fig. 7.19 Tolerancing of a particular part of a feature

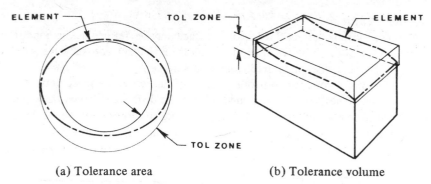

(a) Tolerance area   (b) Tolerance volume

Fig. 7.20 Tolerance zones

Figure 7.20(a) shows a line within its tolerance area and fig. 7.20(b) shows a surface within its tolerance volume.

7.9 Tolerances of form for single features

Tolerances of straightness

The theoretical or perfect straightness of a line on a surface may be defined as the condition in which the distance between any two points on that line is always the shortest possible when measured along the line.

The tolerance zone for controlling errors of straightness is the area between two parallel lines, and the tolerance value is the distance between these lines.

The line on the surface of the feature in fig. 7.21(a) can take any form, provided it lies in an axial plane between two parallel straight lines 0.02 mm apart, as shown in fig. 7.21(b).

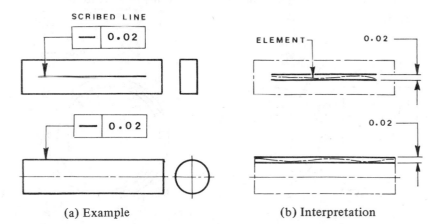

(a) Example   (b) Interpretation

Fig. 7.21 Tolerances of straightness

49

Tolerances of flatness

The theoretical or perfect flatness of a surface may be defined as the condition in which the distance between any two points on that surface is always the shortest possible when measured along that surface.

The tolerance zone for controlling errors of flatness is the space between two parallel planes, and the tolerance value is the distance between these planes.

The surface controlled in fig. 7.22(a) can take any form, provided it lies in the space between two parallel flat planes 0.04 mm apart, as shown in fig. 7.22(b).

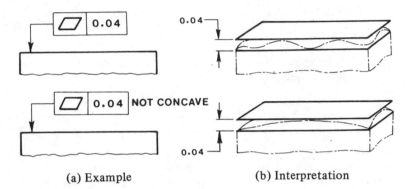

(a) Example   (b) Interpretation

Fig. 7.22 Tolerances of flatness

Tolerances of roundness

The theoretical or perfect roundness of a surface may be defined as the condition in which the surface has the form of a perfect circle, i.e. the distance between any point on the circumference and the centre is always equal to the radius of the circle.

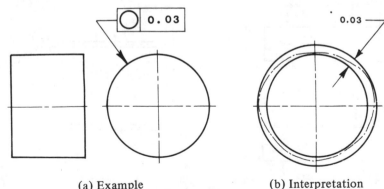

(a) Example   (b) Interpretation

Fig. 7.23 Tolerance of roundness

The tolerance zone for controlling errors of roundness is the annular area between two concentric coplanar circles, and the tolerance value is the radial distance between these circles.

In fig. 7.23(a), the circle controlled, which may represent the periphery at any cross-section perpendicular to the axis, can take any form provided it lies in the space between two concentric circles 0.03 mm radially apart, as shown in fig. 7.23(b).

Tolerances of cylindricity

Theoretical or perfect cylindricity may be defined as the condition in which all cross sections of a solid are perfect circles with their centres lying on a straight axis.

The tolerance zone for controlling errors of cylindricity is the annular space between two perfect cylindrical surfaces lying on the same straight axis, and the tolerance value is the radial distance between these surfaces.

The surface controlled in fig. 7.24(a) may take any form provided it lies between two perfect concentric cylinders 0.03 mm apart, as shown in fig.7.24(b).

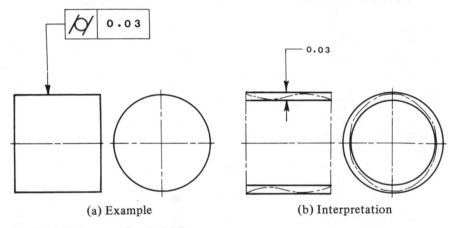

(a) Example   (b) Interpretation

Fig. 7.24 Tolerance of cylindricity

In theory, a cylindricity tolerance could control roundness, straightness, and parallelism; in practice, however, it is difficult to check the combined effect of errors in these characteristics, and it is better to tolerance and inspect each of them separately as required.

Profile tolerance of a line

The theoretical or perfect form of a profile line is defined by boxed dimensions, which locate the true position of any point on that line.

The tolerance zone has a constant width equal to the tolerance value, normal (at 90°) to the theoretical profile and equally disposed about it.

50

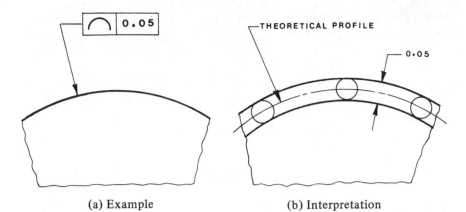

(a) Example (b) Interpretation

Fig. 7.25 Profile tolerance of a line

The tolerance zone is the area between two lines which envelop circles of diameter equal to the tolerance value.

The profile line controlled in fig. 7.25(a) can take any form provided it lies between two lines 0.05 mm apart, as shown in fig. 7.25(b).

Profile tolerances of a surface

The theoretical or perfect form of a surface is defined by boxed dimensions which locate the true position of any point on that surface.

The tolerance zone is the space between two surfaces which envelop spheres of diameter equal to the tolerance value with their centres lying on the theoretical surface of the correct geometrical shape.

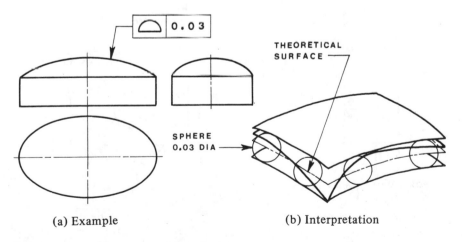

(a) Example (b) Interpretation

Fig. 7.26 Profile tolerance of a surface

The curved surface of the part controlled in fig. 7.26(a) is required to lie between two surfaces as shown in fig. 7.26(b).

7.10 Tolerances of attitude for related features

Tolerances of parallelism

Theoretical or perfect parallelism may be defined as the condition in which all the perpendicular distances between the line or the surface controlled and the datum feature are always the same.

The tolerance zone for controlling errors of parallelism is the area between two parallel straight lines or the space between two parallel planes which are parallel to the datum feature. The tolerance value is the distance between the lines or planes.

The controlled top surface of the part shown in fig. 7.27(a) is required to lie between two planes 0.06 mm apart and parallel to the datum line or surface, as shown in fig. 7.27(b).

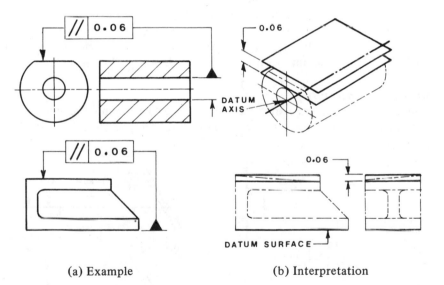

(a) Example (b) Interpretation

Fig. 7.27 Tolerances of parallelism

Tolerances of squareness

Theoretical or perfect squareness may be defined as the condition in which the feature controlled is truly perpendicular to the datum feature.

51

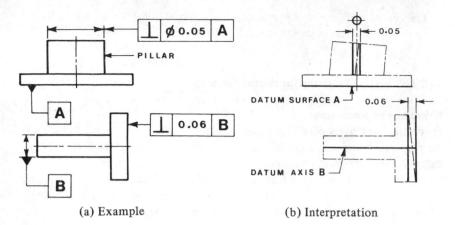

(a) Example   (b) Interpretation

Fig. 7.28 Tolerances of squareness

The axis of the vertical pillar in fig. 7.28(a) is required to be contained within a tolerance cylinder of 0.05 mm diameter, the axis of which is perpendicular to the datum surface A, as shown in fig. 7.28(b).

Note that the tolerance value is here preceded by the symbol $\emptyset$.

The controlled end surface of the second component is required to lie between two planes 0.06 mm apart and perpendicular to the axis of the left-hand cylindrical portion (datum axis B).

Tolerances of angularity

Theoretical or perfect angularity may be defined as the condition in which the controlled feature is inclined to the datum feature at a specified true angle.

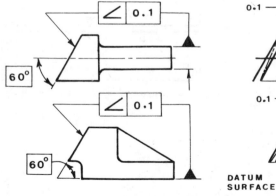

(a) Example   (b) Interpretation

Fig. 7.29 Tolerances of angularity

The tolerance zone for controlling errors of angularity is the area between two parallel straight lines or the space between two parallel planes which are inclined to the datum feature at a specified angle. The tolerance value is the distance separating the lines or planes.

The controlled inclined surfaces of the parts in fig. 7.29(a) are to lie between two planes 0.1 mm apart which are inclined at 60° to the datum axis of the cylindrical portion or the datum surface as shown in fig. 7.29(b).

7.11 Tolerances of location for related features

Tolerances of position

The theoretical position of a feature is the specified true position of the feature as located by boxed dimensions.

The actual point shown in fig. 7.30(a) is required to lie within a tolerance circle 0.1 mm diameter centred on the specified true point of intersection, as shown in fig. 7.30(b).

The axis of the hole is required to be contained within a tolerance cylinder 0.08 mm diameter centred on the specified true position of the axis of the hole.

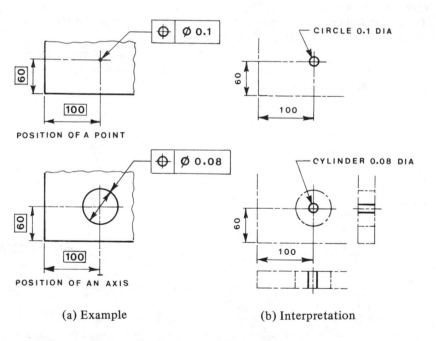

(a) Example   (b) Interpretation

Fig. 7.30 Tolerances of position

Tolerances of concentricity

Theoretical or perfect concentricity may be defined as the condition in which the controlled features (which may be circles or cylinders) lie truly on the same centre or axis as the datum features.

The tolerance zone for controlling errors of concentricity is a circle or cylinder within which the centre or axis of the controlled feature is to be contained. The tolerance value is the diameter of the tolerance zone.

The axis of the right-hand cylindrical portion of the component in fig. 7.31(a) is to be contained within a cylinder 0.08 mm diameter and is to be co-axial with the axis of the left-hand portion, which is the datum, as shown in fig. 7.31(b).

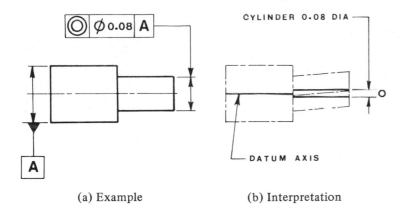

(a) Example   (b) Interpretation

Fig. 7.31 Tolerance of concentricity

Tolerances of symmetry

Theoretical or perfect symmetry may be defined as the condition in which the position of the feature is specified by its perfect symmetrical relationship to a datum.

The tolerance zone for controlling errors of symmetry is the area between two parallel lines or the space between two parallel planes which are symmetrically disposed about the datum feature.

The median plane of the slot controlled in fig. 7.32(a) is required to lie between two parallel planes 0.08 mm apart which are symmetrically disposed about the datum plane, as shown in fig. 7.32(b).

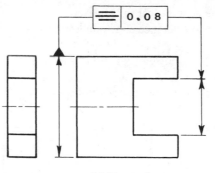

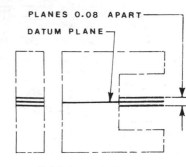

(a) Example   (b) Interpretation

Fig. 7.32 Tolerance of symmetry

7.12 Test questions

1. Define a tolerance.
2. Show by means of neat sketches three possible errors when producing a hole in a component.
3. Name and sketch three main types of engineering fit.
4. Calculate the maximum and minimum limits for the following shaft and hole nominal sizes: (a) 50 H11/c11, (b) 100 H7/n6, (c) 150 H7/s6.
5. With the help of simple sketches, show how limits between the centres of holes can be indicated by (a) chain dimensioning, (b) progressive dimensioning. Identify the main disadvantage of chain dimensioning.
6. The limits between centres of four holes A, B, C, and D are indicated by chain dimensioning in mm:

limits between holes A and B are	20.02
	19.98
limits between holes C and D are	30.02
	29.98
limits between holes A and D are	60.00
	59.88

   Calculate the limits between holes B and C if all holes lie in sequence along the same centre line.
7. Define a geometrical tolerance.
8. State at least three reasons for using geometrical tolerances.
9. Define (a) a tolerance zone, (b) a datum feature.
10. Show how geometrical tolerances can be indicated by means of a rectangular frame.
11. Indicate the difference between form and attitude tolerances.
12. Name and sketch the symbols for (a) three form tolerances, (b) two attitude tolerances, (c) two location tolerances.

13. State at least three advantages of using geometrical tolerances.
14. Show two methods of how a geometrical tolerance controlling the top surface in fig. 7.33(a) can be indicated.

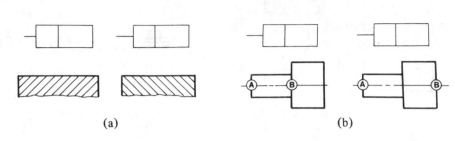

(a)   (b)

Fig. 7.33

15. If the geometrical tolerance refers to the axis between A and B, complete fig. 7.33(b) and indicate the position of the arrowheads.
16. Show two methods of how the top surface of the component in fig. 7.34(a) can be indicated as the datum.

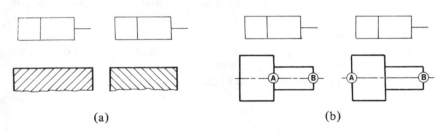

(a)   (b)

Fig. 7.34

17. If the axis between A and B is the required datum feature, show how this can be indicated in fig. 7.34(b).
18. Complete the drawings in fig. 7.35 by indicating the correct geometrical tolerances to satisfy the following conditions in each case: (a) the axis of the whole component is required to be contained in a cylindrical zone 0.03 mm diameter; (b) the top surface of the component is required to lie between two parallel planes 0.03 mm apart; (c) the periphery at any cross-section perpendicular to the axis is required to lie between two concentric circles 0.03 mm radially apart; (d) the right-hand face of the component is required to lie between two parallel planes 0.03 mm apart and perpendicular to the top surface.
Tracing paper may be used.

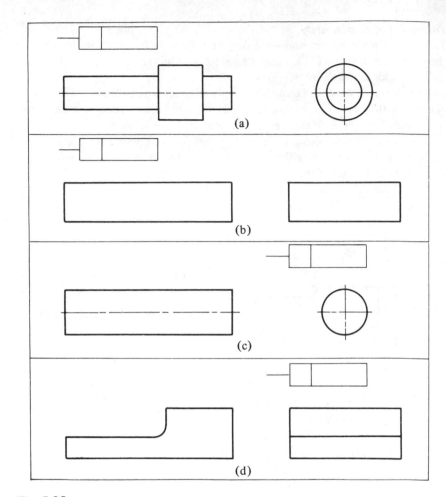

(a)

(b)

(c)

(d)

Fig. 7.35

19. Complete the drawings in fig. 7.35 by indicating the correct geometrical tolerances to satisfy the following conditions in each case: (a) the axes of the right-hand and left-hand cylindrical portions are required to be contained within one cylinder 0.02 mm diameter; (b) the top face of the part is required to lie between two parallel planes 0.08 mm apart which are perpendicular to the datum plane, which is the right-hand face; (c) the curved surface of the part is required to lie between two cylindrical surfaces co-axial with each other, a radial distance of 0.03 mm apart; (d) the top surface of the part is required to lie between two planes 0.05 mm apart and parallel to the datum plane.
Tracing paper may be used.

54

20. By means of neat sketches and explanatory notes, interpret the geometrical tolerances in fig. 7.36. Tracing paper may be used.

21. Name and interpret by means of neat sketches and explanatory notes the geometrical tolerances shown in fig. 7.37. Tracing paper may be used.

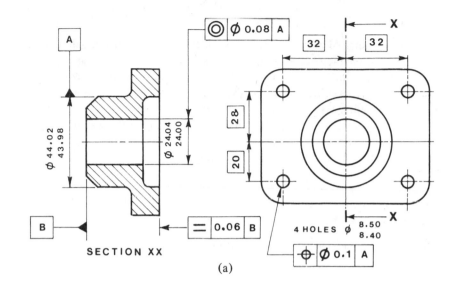

(a)

(b)

Fig. 7.37

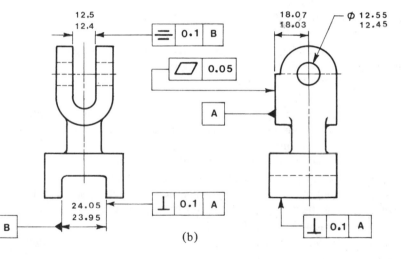

Fig. 7.36 (a) (b) (c) (d) (e) (f)

8 Basic engineering materials

There is a vast range of modern materials available to the engineering designer, who faces a formidable task when selecting the most suitable material for a particular component to perform a specific function, as there are so many different factors that might influence his choice. These factors include the design requirements, mass, loading, service life, climatic and chemical environment, reliability, machining requirements, manufacturing processes, properties of materials, and considerations of cost, quantity required, etc.

The properties of materials can be subdivided into:

a) Physical and chemical properties These include melting point, thermal conductivity, electrical conductivity, density, coefficient of linear expansion, corrosion resistance, etc.

b) Mechanical properties Some of these properties are discussed below.

Strength is the ability of a material to resist applied forces without fracture or permanent deformation. The applied forces may produce tension, compression, shear, or torsion. Impurities in the material, hot or cold working, heat treatment, and other factors can affect the strength of a material.

Elasticity is the ability of a material to return to its original shape after being deformed or strained due to stress resulting from the application of forces.

Ductility is the ability of a material to deform considerably under a tensile load before failure. Materials having this property can be formed into various shapes by bending, drawing, extruding, rolling, etc. Those possessing high ductility are gold, aluminium, silver, lead, and certain grades of steel and brass.

Brittleness is the property of a material of fracturing without much permanent distortion, especially when subjected to suddenly applied forces. Brittleness is the lack of ductility. Materials possessing brittleness are most cast irons, glass, and certain grades of steel, especially when the phosphorus content is too high.

Malleability is the ability of a material to be permanently deformed in compression without fracture; for example by hammering, forging, pressing, rolling, etc. This property is very similar to ductility. Materials posessing malleability are lead, tin, zinc, and gold.

Toughness is the ability of a material to resist fracture, especially under suddenly applied forces. It is the reverse of brittleness. Tough materials are capable of absorbing a large amount of impact energy without fracture and some can be repeatedly bent or twisted. Materials possessing toughness are wrought iron, aluminium, copper alloys, etc.

Hardness is the ability of a material to resist scratching, wear, abrasion, and indentation. The hardness of a material can be improved by alloying, or by a suitable heat treatment. Materials possessing high hardness include diamond, tungsten carbides, ceramics, heat-treated alloy cast irons, and some steels.

Machinability is the ability of a material to be machined with ease.

In general, engineering materials belong to two main groups — metals and non-metals — as shown in fig. 8.1. Tables 8.1, 8.2, and 8.3 show the properties and indicate the typical uses of some engineering materials. These tables should serve as an approximate guide for selection of suitable materials for engineering components.

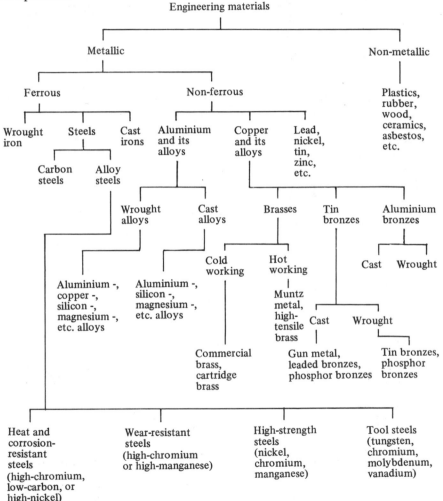

Fig. 8.1 Basic engineering materials

Symbols and melting points (°C) of some elements

Aluminium (Al), 660	Carbon (C)	Chromium (Cr), 1900	Lead (Pb), 327	Manganese (Mn), 1244	Nickel (Ni), 1453	Silicon (Si), 1412 Tin (Sn), 232
Antimony (Sb), 630	Copper (Cu), 1083	Iron (Fe), 1536	Magnesium (Mg), 650	Molybdenum (Mo), 2620 Phosphorus (P), 44	Sulphur (S), 113	Zinc (Zn), 420

Table 8.1 Properties and uses of irons, plain-carbon steels, and alloy steels. (These materials are generally cheap, widely available, and strong, with a wide range of properties.)

Material and approximate composition (%)	Relative density	Condition	Elasticity	Tensile strength (MPa)	Ductility	Hardness	Toughness	Machin-ability	Additional properties	Typical uses
Wrought iron Fe (almost pure)	7.8	Hot-rolled	Medium	340	Medium–high	Medium	High	Good	Malleable; shock resistant; can be bent, forged, hammer-welded	Chain links, ornamental work etc.
Grey cast irons (3–4)C, (1.2–2.8)Si, (0.5–1)Mn	7.2	Cast	Extremely low	170–350	Approx. zero	Very high	Very low	Good	Cheap; easily cast; corrosion & wear resistant; strong in compression; damps vibrations; 'self-lubricating'	Brackets, machine frames, pistons, cylinders, pipes, pulleys, gears, bearings, slides, etc.
Malleable irons (2–3)C, (0.1–0.5)P, (0.5–6)S, (1–5)Si, (0.4–21)Ni, (0.1–5)Cr	7.2	Cast Heat-treated	Medium	280–510	Very low–low	Medium	Low	Excellent–good	Increased ductility & malleability; shock & corrosion resistant; not weldable; can be cast and forged	Brake drums, levers, links, shafts, hinges, spanners, chains, wheels, vice bodies, etc.
Spheroidal-graphite (SG) irons Nodular irons	7.3	Cast	Very high	370–725	Medium	Very low–high	Medium	Fair–excellent	Added magnesium reduces graphite flakes to spheroids, increasing ductility, strength & shock resistance	Machine frames, pump bodies, pipes, crankshafts, hand tools, gears, dies, office equipment, etc.
Low-carbon steels (up to 0.25)C	7.85	Cannot be heat-treated	High	430–480	Medium	Medium	High	Fair–good	Most commonly used; cheap; magnetic; can be welded, forged, case-hardened	Lightly stressed parts, nails, car bodies, chains, rivets, wire, structural parts, etc.
Medium-carbon steels (0.25–0.6)C	7.85	Heat-treated	High–very high	480–620	Low–medium	High	Medium	Fair–poor	Weldable at lower carbon contents. With increased carbon contents brittleness is increased.	Axles, spindles, couplings, shafts, tubes, gears, forgings, rails, hand tools, dies, ropes, keys, etc.
High-carbon (tool) steels (0.6–1.5)C	7.85	Heat-treated	Very high–extremely high	620–820	Low	Very high	Very low	Very poor (anneal)	With increased carbon content brittleness is further increased & ductility is decreased	Hammers, chisels, screws, drills, taps, dies, blades, punches, knives, chisels, saws, razors, etc.
Nickel steels (0.1–4)C, (0.04–1.5)Mn, (1–5)Ni	7.8	Heat-treated	Medium–extremely high	310–700	Medium	High	Very high	Poor	Nickel improves strength and toughness	Axles, crankshafts, car parts, camshafts, gears, pins, pinions, etc.
Stainless steels (0.05–0.1)C, (0.8–1.5)Mn, (8.5–18)Ni, (12.5–18)Cr	7.9	Heat-treated	Medium–very high	650–900	Medium–high	Medium–very high	Medium–low	Good–fair	Over 12% chromium protects surfaces from corrosion 18/8 (Cr/Ni) steel is acid resistant.	Chemical plants, kitchen equipment, cutlery, springs, circlips, etc.
Low-alloy nickel-chrome steels (1–5)Ni, (0.6–1.5)Cr	7.8	Heat-treated	High–extremely high	930–1500	Low	Very high	High	Good	With heat treatment, a wide range of properties may be obtained.	Highly stressed parts: con-rods, shafts, gears, driving shafts, crankshafts, etc.
Manganese steels (0.35–1.2)C, (1.5–12.5)Mn	7.9	Heat-treated	Very high	700–850	Medium–high	Very high	High	Extreme-ly poor	Wear resistant; non-magnetic	Cutting tools, stone-crushing jaws, dredging equipment, press tools, railway crossings, etc.
Heat-resisting steel (0.1)C, (1.5)Si, (1)Mn, (19)Cr, (11)Ni	7.9	Air-cooled	High	690	High	Medium	High	Poor	Resistant to heat & thermal shock; 1000°C max. working temperature	Components exposed to high temperatures etc.

Table 8.2 Properties and uses of aluminium alloys and copper alloys (These materials generally have pleasing appearance, with corrosion-resisting properties.)

Material and approximate composition (%)	Relative density	Condition	Elasticity	Tensile strength (MPa)	Ductility	Hardness	Machin-ability	Additional properties	Typical uses
Aluminium Al (almost pure)	2.7	Annealed Hard	Very low–low	55–140	High–low	Very low–low	Fair	Good electricity & heat conductor; non-magnetic; malleable; can be forged & extruded; non-corrosive	Electrical cables, reflectors, cooking utensils, radiators, piping, building components, paints, etc.
Aluminium–silicon alloy (88)Al, (12)Si, + traces of Fe & Mn	2.7	Cast	Low	280–140	Extremely low	Very low	Fair	High castability and corrosion resistance; can be pressure die-cast	Light castings, aircraft & marine applications, radiators, crankcases, gearboxes, etc.
Duralumin (4)Cu, (0.8)Mg, (0.5)Si, (0.7)Mn	2.8	Heat-treated	High–very high	450–550	Low	High	Good	Very high strength/weight ratio	General purposes, stressed aircraft components, structural components, etc.
Copper Cu (almost pure)	8.9	Annealed Hard	Very low–very high	220–350	Extremely high – very low	Very low–medium	Poor	Good heat & electrical conductor; corrosion resistant; easily brazed; can be drawn & forged	Chemical industry, heating equipment, cooking utensils, tubing, roofing, boilers, etc.
Cartridge brass (70)Cu, (30)Zn	8.5	Annealed Hard	Low–very high	325–650	Very high–low	Low–high	Poor	Non-magnetic; corrosion resistant; can be forged, drawn extruded	Cartridge, shells, jewellery, etc.
Yellow brass or Muntz metal (60)Cu, (40)Zn	8.3	Annealed Hard	Low–high	355–465	High–low	Medium–very high	Fair	Non-magnetic; corrosion resistant; poor forming qualities	Structural plates, tubing, valve rods, hot forgings, etc.
Commercial brass (90)Cu, (10)Zn	8.8	Annealed Hard	Very low–very high	280–510	Very high–very low	Low–high	Good	Can be drawn, bent, brazed, cold worked, welded, enamelled	Imitation jewellery, lipstick cases, clamps, etc.
Tin bronze (89)Cu, (11)Sn	8.7	Annealed Hard	Low	220–310	Low–very low	Low–medium	Fair	Can be cast	Bearings, bushes, gears, piston rings, pump bodies, etc.
Gun-metal (tin bronze) (88)Cu, (10)Sn, (2)Zn	8.5	Annealed Hard	Medium–very high	270–340	Medium–low	Low–medium	Fair	Good castability; resistant to salt-water corrosion	Bearings, steam valve bodies, marine castings, structural parts, etc.
Aluminium bronze (91–95)Cu, (5–9)Al	7.6	Annealed Hard	Low–very high	370–770	Very high–low	Low–very high	Very poor	Corrosion & heat resistant; sea-water resistant; can be welded; difficult to machine	High wear and strength applications, marine hardware, etc.
Phosphor bronzes (86–90.7)Cu, (9–13)Sn, (0.3–1)P	8.9	Cast	Low–very high	220–420	High–very low	Low–very high	Fair	Phosphorus improves tensile strength; wear and corrosion resistant	Bearings, bushes, valves, general sand castings, etc.
High-lead tin bronze (76)Cu, (9)Sn, (15)Pb	9.1	Cast	Very low–medium	170–310	Medium–very low	Low–medium	Very good	Resistant to acid corrosion	General-purpose bearing and bushing alloy, wedges, etc.
Monel metal (30)Cu, (1.4)Fe, (1)Mn, (67.6)Ni	8.8	Annealed Hard	High–extremely high	600–950	High–low	High–very high	Fair	Very high corrosion resistance; can be cast, forged, stamped, & drawn; heat resistant	Chemical engineering, propeller shafts, high-temperature valve seats, high-strength components, etc.

Table 8.3 Properties and uses of plastics (These materials are very light, easily formed, sometimes transparent, corrosion resistant, with excellent electrical resisting properties.)

Group	Compound	Relative density	Flamm-ability	Tensile strength (MPa)	Maximum working temperature (°C)	Chemical resistance	Relative cost	Additional properties	Typical uses
Thermoplastic materials. (These materials can be repeatedly softened and moulded by heating, then hardened by cooling.)									
Cellulosics	Cellulose nitrate (Celluloid)	1.37	Flammable	50	55	Fair	Low	Tough; dimensionally stable; low water absorption; inflam-mable; sensitive to heat	Handles, piano keys, toilet seats, fountain pens, spectacle frames, instrument labels, etc.
	Cellulose acetate (Tricel)	1.22–1.31	Slow burning	20–60	70	Fair	Low	Hard; stiff; strong; tough; transparent; surfaces of high gloss can be obtained.	Photographic film, artificial leather, lamp shades, toys, combs, cable covering, tool handles, etc.
Vinyls	Polyvinyl chloride (Rigid PVC) (Plasticized PVC)	1.34–1.40	Self-extinguishing	50	70–105	Good	Moderate	Rigid PVC is hard, tough, strong, stiff, abrasion resistant, Plasticized PVC is more flexible.	Pipes, bottles, chemical plant, lighting fittings, curtain rails, cable covers, toys, balls, gloves, etc.
	Polypropylene	0.9	Slow burning	35	100	Very good	Moderate	Strong; stiff; very light; good temperature resistance & electrical insulation properties	Pipes & fittings, bottles, crates, cable insulation, tanks, cabinets for radios, shoe heels, pumps, etc.
	Polystyrene	1.05–1.08	Slow burning	50	100	Good	Low	Rather brittle; transparent; Toughening with rubber improves impact & heat resistance	Vending machine cups, housings for cleaners & cameras, radio cabinets, furniture, toys, etc.
Fluorocarbons	Polytetrafluorethylene (PTFE, 'Teflon', etc.)	2.13–2.19	Non-flammable	21	260	Excellent	Very high	Low coefficient of friction; tough; heat resistant; weather & corrosion resistant; can be machined	Bearings, gaskets, valves, chemical plant, electrical-insulation tapes, non-stick coatings for frying pans
Polyamides	Nylon 66	1.14	Self-extinguishing	70	150	Good	High	Stiff; strong; tough; abrasion resistant; high degree of rigidity; low-friction property	Bearings, gears, cams, pulleys, combs, bristles for brushes, ropes, fishing lines, raincoats, containers
Acrylics	Polymethylmethacrylate ('Perspex', 'Plexiglas', etc.)	1.19	Slow burning	55–80	93	Good	Moderate	Completely transparent; strong; stiff; shatter & weather resistant; will not discolour; can be decorated	Lenses, aircraft glazing, windows, roof lighting, sinks, baths, knobs, telephones, dentures, machine guards
Thermosetting materials (These materials undergo a chemical change when moulded: they become permanently rigid and incapable of being softened again.)									
Phenolics	Phenol formaldehyde ('Bakelite', etc.)	1.35–1.5	Very slow burning	50–60	120	Very good	Low	Brittle; heavy; hard; rigid; dark coloured; darken under influence of light; very popular thermoset	Vacuum cleaners, ashtrays, buttons, cameras, electrical equipment, dies, handles, gears, costume jewellery
	Melamine and Urea formaldehyde	1.4–1.55	Very slow burning to non-flammable	45–75	100	Good	Moderate	Available in light colours; good colour stability; brittle; heavy; hard; rigid	Electrical equipment, handles, cups, plates, trays, radio cabinets, knobs, building panels, etc.
Epoxides	Epoxy resins ('Bakelite', 'Araldite', etc.)	1.12–1.19	Slow burning to self-extinguishing	60	170	Good	Moderate	Due to many constituents, epoxides can be liquids, solutions, pastes, or solids	Adhesives, surface coatings, flooring, electrical insulation glass-fibre laminates, furniture, etc.
High-pressure laminates	Laminates ('Tufnol', 'Formica', etc.)	1.15–1.75	Slow burning to self-extinguishing	12–80	230–300	Good	High–low	Layers of paper, fabric, etc. bonded under pressure with a resin. Extremely strong; readily punched & machined; wear resistant	(Paper l's) electrical insulation. (Fabric l's) gears, bearings, jigs, aircraft parts, press tools. (Decorative l's) table tops, trays.

9 Methods of fastening components

Most engineering assemblies require some sort of fastenings for joining two or more components together. These fastenings may be temporary or permanent.

Temporary fastenings can be used more than once. Any assembled components held together can be dismantled and re-assembled together many times without damaging the fastenings. They include screws, bolts, studs, pins, keys, etc.

Permanent fastenings have to be damaged or destroyed, if the components need to be separated after they have been joined. They include rivets, adhesives, welding, soldering, etc.

9.1 Threaded fastenings

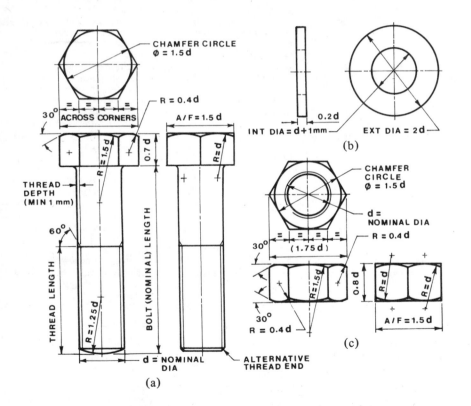

Fig. 9.1 (a) General proportions of (a) bolt, (b) washer, and (c) nut

Nuts and bolts

The bolt has a plain portion of a shank between the head and the start of the thread. Bolts generally pass completely through the components to be fastened and are secured by a nut on the other side.

Design proportions for drawing bolts, nuts and washers are shown in fig. 9.1. Students must try to memorise all these proportions, which are necessary for drawing more complicated assemblies. If an approximate method of drawing nuts and bolts is required, reference should be made to *Engineering drawing for technicians volume 1.*

Studs

Studs are used for components that are removed frequently, like cylinder heads, covers, lids, etc. The stud consists of the following parts:

a) The *metal end,* which is screwed fully into one of the components. The thread length should be equal to *d* (the nominal diameter of the thread) or 1.5*d*.
b) The *nut end,* which accommodates a nut. (The thread length is indicated in Table 9.1.)
c) The *plain portion,* which is the unthreaded length of stud and should not be less than 0.5*d*. This part accommodates the other component.

The nominal length of the stud is measured from the nut end, including the chamfer, to the extreme end of the run-out of the thread on the metal end.

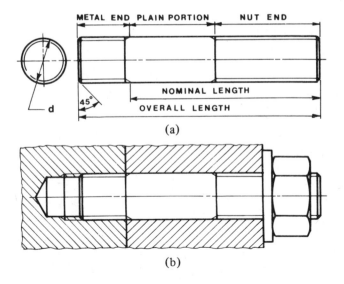

Fig. 9.2 (a) Stud and (b) stud assembly

Table 9.1 Standard thread lengths for bolts and nut-end lengths for studs

Nominal length of bolt or stud	Length of thread
Up to and including 125 mm	$2d + 6$ mm
Over 125 mm and up to and including 200 mm	$2d + 12$ mm
Over 200 mm	$2d + 25$ mm

Screws

A machine screw is threaded for the entire length of the shank.

Figure 9.3(a) shows the hexagon-head screw. The other types of screw are shown in fig. 9.4.

In order to facilitate tapping of a blind tapped hole, the depth of the drilled hole should always be greater than the length of the threaded portion, as shown in figs 9.2(b) and 9.3(a).

To ensure that the nut is fully engaged when assembled, the bolt or stud thread must be long enough to extend on both sides of the nut in its tightened position, as shown in figs 9.2(b) and 9.3(b).

Normally only one washer is used for an assembly of a bolt, stud, or screw. The washer should be placed under the component which is turned in order to tighten the assembly, as shown in figs 9.2(b) and 9.3(a) and (b).

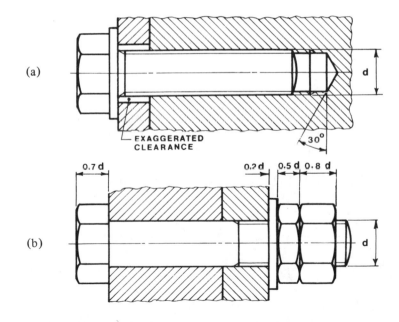

Fig. 9.3 (a) Screw assembly, (b) bolt assembly with a lock nut

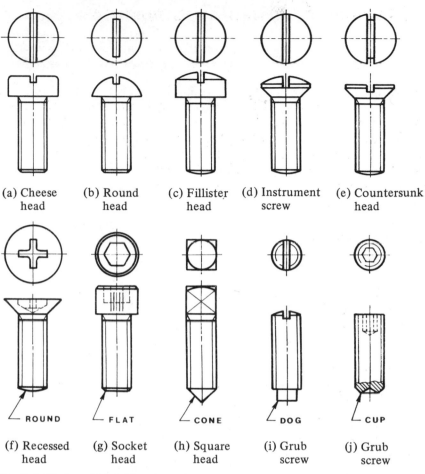

(a) Cheese head (b) Round head (c) Fillister head (d) Instrument screw (e) Countersunk head

(f) Recessed head (g) Socket head (h) Square head (i) Grub screw (j) Grub screw

Fig. 9.4 Various types of screw

Clearance holes to accommodate bolts, studs, and screws are not normally drawn. If it is necessary to draw them, they should be shown with the clearance exaggerated, as shown in fig. 9.3(a).

Bolts, studs, nuts, screws, and washers are not, by convention, to be sectioned longitudinally.

Cheese-head (a), *round-head* (b) and *fillister-head* (c) screws (fig. 9.4) are used regularly in engineering, whereas *instrument screws* (d) are used in instrument work. *Countersunk-head screws* (e) are used where a flush surface has to be maintained. *Recess-head screws* (f) need a special cross-shaped screwdriver for tightening. *Socket-head screws* (g) can be placed in counterbored holes to maintain a flush surface. They are tightened by means of a hexagonal wrench and are mainly used in tool-making. *Square-head set screws* (h) and headless

61

grub screws (i) and (j) are used to prevent relative rotation or sliding movement between two components.

Flat pointed (g) or *cup-pointed* (j) screws are used where a contact with flat surfaces is required to prevent movement. *Cone-pointed screws* (h) bite into shafts, and *dog-pointed screws* (i) usually fit into slots.

Self-tapping screws (not shown) form their own thread when tightened.

9.2 Keys

A key is a component inserted between the shaft and the hub of a pulley, wheel, etc., to prevent relative rotation but allow sliding movement along the shaft, if required.

The recess machined in a shaft or hub to accommodate the key is called a *keyway*. Keyways can be milled horizontally or vertically, as shown in fig. 9.5. Keys are made of steel, in order to withstand the considerable shear and compressive stresses caused by the torque they transmit.

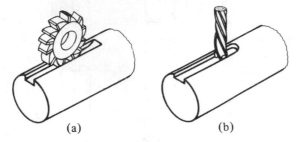

Fig. 9.5 Keyways milled (a) horizontally, (b) vertically

There are two basic types of key:
a) *Saddle keys*, which are sunk into the hub only. These keys are suitable only for light duty, since they rely on a friction drive alone.
b) *Sunk keys*, which are sunk into the shaft and into the hub for half their thickness in each. These keys are suitable for heavy duty, since they rely on positive drive.

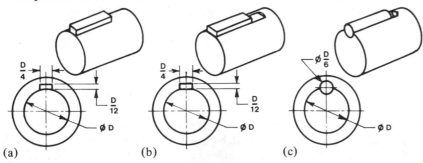

Fig. 9.6 (a) Hollow saddle key, (b) flat saddle key, and (c) round key

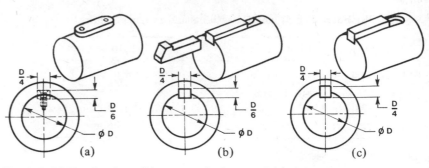

Fig. 9.7 (a) Feather key, (b) rectangular key, and (c) square key

Hollow saddle keys (fig. 9.6(a)) are used for very light duty.
Flat saddle keys (fig. 9.6(b)) are used for light duty.
Round keys (fig. 9.6(c)) are used for medium duty.

A *feather key* is used when the hub is required to slide along the shaft. It is tightly fitted or secured by means of screws in the shaft keyway, and is made to slide in the hub keyway, as shown in fig. 9.7(a).

Rectangular and square keys can be parallel or tapered with a basic taper of 1 in 100 to prevent sliding. These keys are used for heavy-duty applications. Students are advised to use square keys for assembly-drawing solutions. *Gib heads* are sometimes provided on taper keys to facilitate their withdrawal, as shown in fig. 9.7(b).

A *Woodruff key* is an almost semi-circular disc which fits into a circular keyway in the shaft. The top part of the key stands proud of the shaft and fits into the keyway in the parallel or tapered hub, as shown in fig. 9.8(a). As the key can rotate in the keyway, it can fit any tapered hole in a hub.

A *splined shaft* is used when the hub is required to slide along the shaft, as shown in fig. 9.8(b). These shafts are used mostly for sliding-gear applications. The splines are usually milled and the splined holes broached.

Square-head set screws and *grub screws* are also used for low-torque applications as shown in fig. 9.8(c) and also 9.4(h), (i), and (j).

If the torque to be transmitted is too great for one grubscrew or key, two may be used set at 90° to 120° around the shaft, but never at 180°.

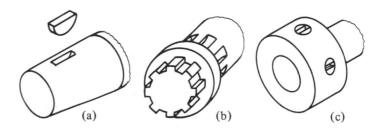

Fig. 9.8 (a) Woodruff key, (b) splined shaft, and (c) grub screws

10 Engineering drawing

10.1 Single-part or detail drawings

These working drawings usually show a single component and should give all the information necessary for the manufacture of the component.

Some of the items which a working drawing should specify are:

a) the form of the component,

b) the full dimensions and tolerances,

c) the material to be used and its specifications,

d) heat-treatment/hardness instructions,

e) the surface texture and finish required,

f) manufacturing processes, machining instructions, etc.

Title blocks

So that any drawing may be stored and, when required, be identified and located quickly, an efficient system of labelling and cross-referencing is required. To facilitate this, all drawings must have a title block, which should contain the following information required for identification and interpretation of the drawing:

a) the name of firm (or college),

b) the drawing number,

c) the title,

d) the scale ratio used,

e) the date of the drawing.

f) the signature of the draughtsman (or student),

g) the projection symbol (first- or third-angle),

h) the units of measurement used.

and additionally:

j) the material used and its specification,

k) the heat treatment/hardness required,

l) general tolerances,

m) warning notes, e.g. 'DO NOT SCALE', etc.

Arrangements and positioning of title blocks differ considerably, but many drawing offices use sheets which are bought already printed and are of a standard size and layout.

The title block should preferably be positioned at the bottom of the sheet, with the drawing number in the lower right-hand corner. For filing reference purposes, the drawing number may also appear in the top left-hand corner of the drawing, as shown in figs 10.1 and 10.3.

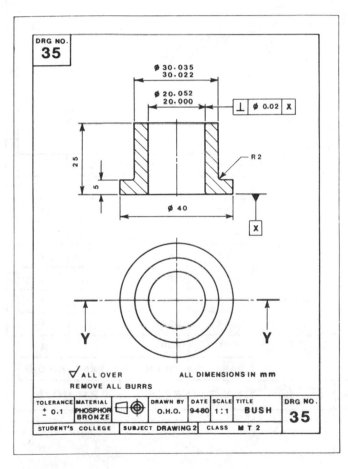

Fig. 10.1 Single-part or detail drawing

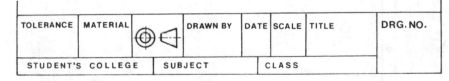

TOLERANCE	MATERIAL			DRAWN BY	DATE	SCALE	TITLE	DRG. NO.
STUDENT'S COLLEGE		SUBJECT			CLASS			

Fig. 10.2 Title block for college use

10.2 Assembly drawings

When a large machine is designed, such as a lathe, an assembly drawing is prepared to show the general arrangement of the machine. This drawing should show the entire finished product with all parts assembled in their correct relative positions.

63

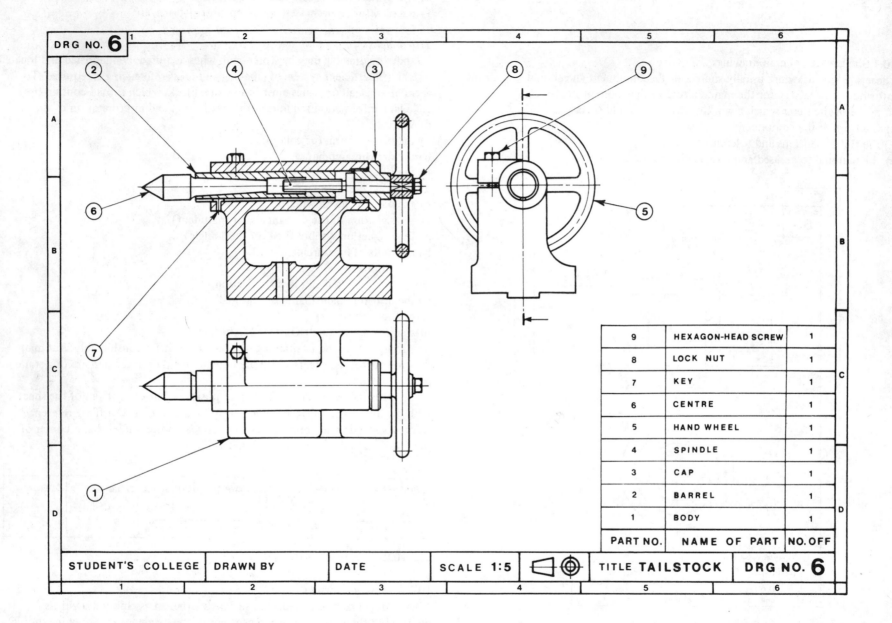

PART NO.	NAME OF PART	NO. OFF
9	HEXAGON-HEAD SCREW	1
8	LOCK NUT	1
7	KEY	1
6	CENTRE	1
5	HAND WHEEL	1
4	SPINDLE	1
3	CAP	1
2	BARREL	1
1	BODY	1

STUDENT'S COLLEGE · DRAWN BY · DATE · SCALE 1:5 · TITLE **TAILSTOCK** · DRG NO. **6**

Fig. 10.3 Sub-assembly drawing

An assembly drawing may include overall dimensions and functional and fitting dimensions.

Sub-assembly drawings

As the machine is large and consists of many small parts, it is necessary to prepare additional sub-assembly drawings of its main parts, such as the bed, headstock, tailstock, carriage, etc. of a lathe — see fig. 10.3.

Parts lists

When all parts in an assembly drawing have to be identified, each single part is usually labelled by means of a reference number, which may be its detail-drawing number or an independent item number.

The separate parts comprising the assembly are located in the drawing by leaders radiating from the circles, or 'balloons', which contain the relevant reference numbers and are usually listed in a parts list. For small assemblies the parts list is placed next to the title block on the drawing; for large assemblies it is usually on a sheet separate from the drawing.

A typical parts list might include the following, as shown in fig. 10.4:
a) the part number,
b) the name or description of the part,
c) the material from which the part is to be made,
d) the quantity required.

PART NO.	NAME OF PART	MATERIAL	NO.OFF
3			
2			
1			

Fig. 10.4 Parts list

Grid system or zoning

Assembly drawings may include a grid-reference system which is based on numbered and lettered divisions in the margin of the drawing sheet (see fig. 10.3). This facilitates the quick location of a part, particular feature, or certain dimensions.

This grid-reference system may also be used to locate and identify subsequent amendments to the drawing. These should be entered in a revision table with the appropriate zone reference, e.g. C5, D3, etc.

10.3 Auxiliary views

Occasionally a component has surfaces which are not parallel with any of the principal planes of orthographic projection and which therefore cannot be clearly defined or dimensioned. To draw the true shapes of those surfaces, additional views are required showing the surfaces as they appear when looking directly at them. These views are called *auxiliary views*.

Figure 10.5 shows an object suspended inside a third-angle-projection 'glass box' which consists of three principal planes and two auxiliary planes.

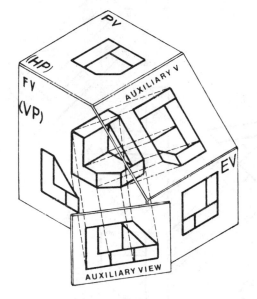

Fig. 10.5 Principal and auxiliary planes

When viewing the object directly, at 90°, through the vertical principal plane (VP), the image of the object will correspond to the front view (FV). When this vertical plane is rotated through 90° about a vertical axis, the corresponding view is the end view (EV). Any other position of the vertical plane between the front view and the end view will correspond to an auxiliary view.

Similarly, when the vertical plane is rotated through 90° about a horizontal axis, it becomes the horizontal principal plane (HP) and its corresponding view is the plan view (PV). Again, any intermediate position of the plane is an auxiliary plane with its corresponding auxiliary view.

In general, referring to fig. 10.5, any view other than the front, plan, or end view is an auxiliary view.

Figure 10.6 shows how the auxiliary view is obtained between the front and end views in third- and first-angle projection.

Figure 10.7 shows how the auxiliary view is obtained between the end and plan views in third- and first-angle projection.

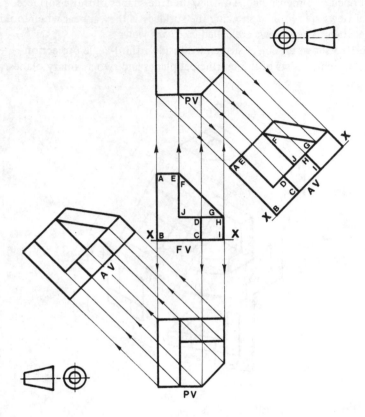

Fig. 10.6 Auxiliary view

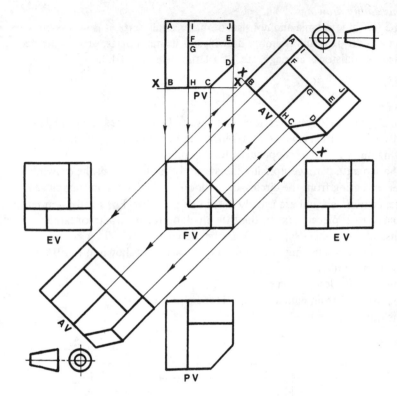

Fig. 10.7 Auxiliary view

To draw the auxiliary view (fig. 10.6)
1. Draw the front view (FV) and plan view (PV).
2. Draw the chosen reference line XX in the front view and similarly in the auxiliary view at right angles to the direction of viewing.
3. Transfer all required vertical distances measured from XX in the front view along projectors to the relevant points in the plan view and reflect them to the auxiliary view. Measure the same distances from the reference line XX in the auxiliary view.
4. Join the points so obtained to complete the required auxiliary view (AV).

To draw the auxiliary view (fig. 10.7)
1. Draw the front and plan views.
2. Draw the chosen reference line XX in the plan view and similarly in the auxiliary view at right angles to the direction of viewing.
3. Transfer all required vertical distances measured from XX in the plan view along projectors to the relevant points in the front view and reflect them to the auxiliary view. Measure the same distances from the reference line XX in the auxiliary view.
4. Join the points so obtained to complete the required auxiliary view (AV).

10.4 Test questions

1. Draw an auxiliary view of each component shown in fig. 10.8, looking in the direction of the arrow normal to the inclined surface. Components 1 to 4 are in first-angle projection and 4 to 6 in third-angle. Tracing paper may be used.

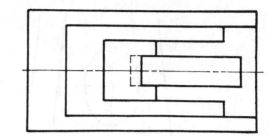

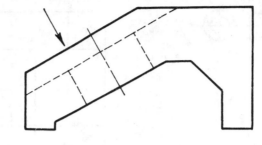

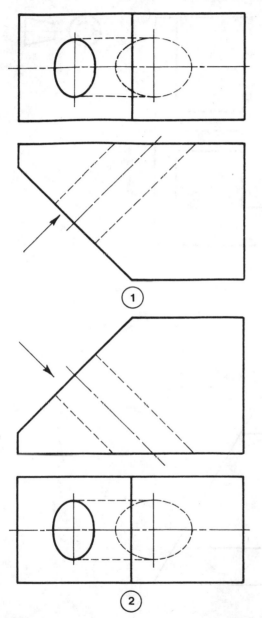

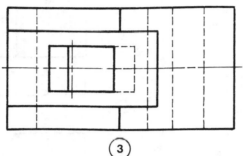

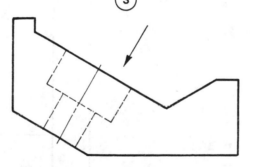

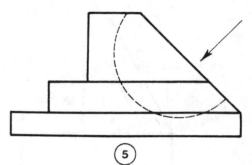

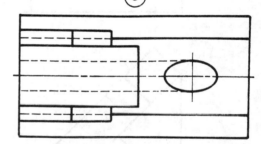

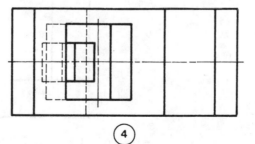

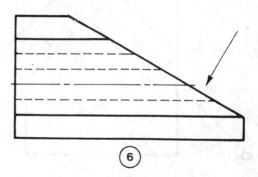

Fig. 10.8

2. For the component shown in fig. 10.9, draw an auxiliary view looking in the direction of the arrow B and an end view looking in the direction of the arrow A. Tracing paper may be used.

3. Draw an auxiliary view of the component shown in fig. 10.10, looking in the direction of arrow A. Tracing paper may be used.

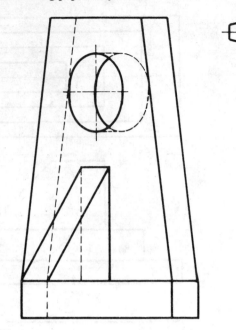

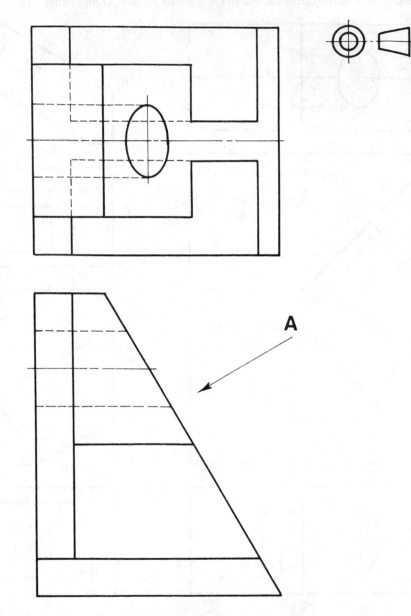

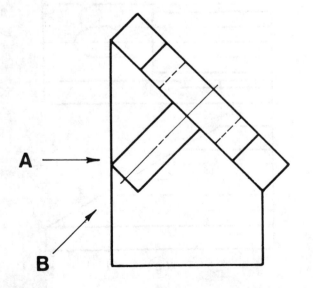

Fig. 10.9

Fig. 10.10

68

4. Figure 10.11 shows components in first-angle projection. Draw an auxiliary view of each component, looking in the direction of the arrow normal to the inclined surface. Tracing paper may be used.

5. Figure 10.12 shows components in third-angle projection. Draw an auxiliary view of each component, looking in the direction of the arrow normal to the inclined surface. Tracing paper may be used.

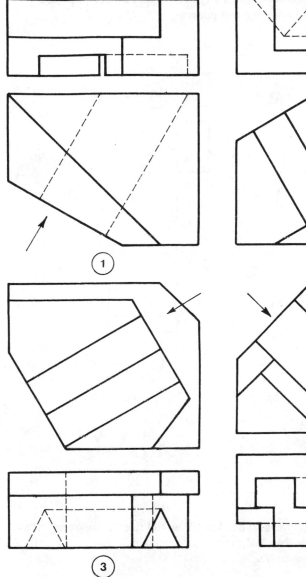
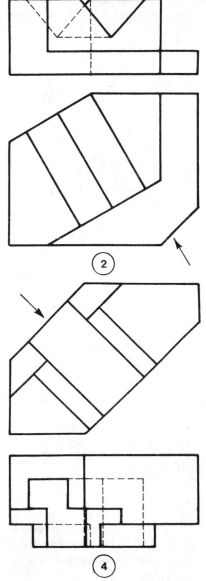

Fig. 10.11

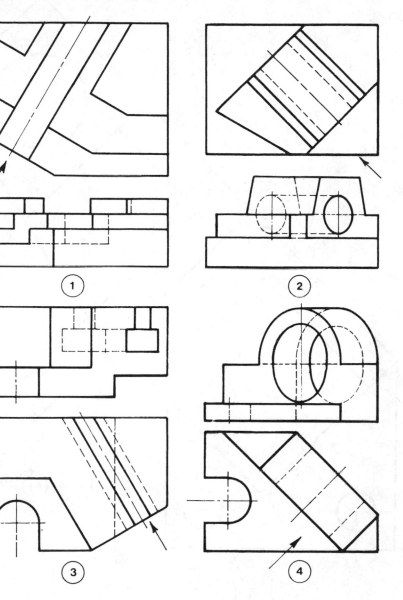

Fig. 10.12

6. Draw an auxiliary view of the component shown in fig. 10.13, looking in the direction of arrow A. Tracing paper may be used.
7. Explain the difference between assembly and sub-assembly drawings.
8. List six items of information that a typical block should contain.
9. List five items of information which a typical single-part drawing should specify for the manufacture of a component.

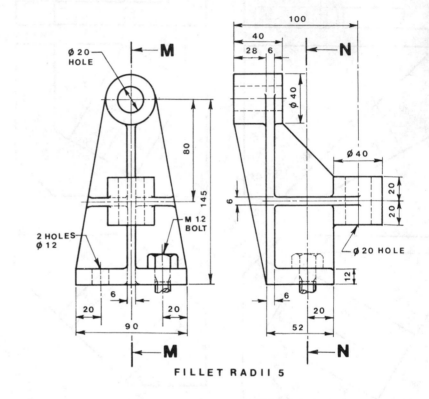

Fig. 10.14

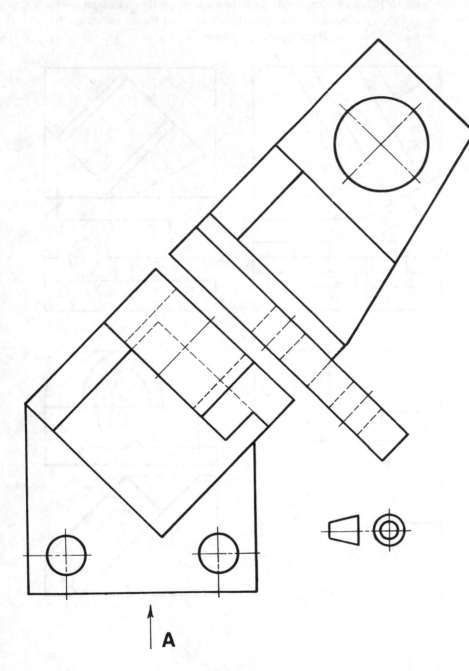

Fig. 10.13

10. Figure 10.14 shows a bevel-gear bracket. Draw full size the following views:
    a) a sectional front view on MM,
    b) a sectional end view on NN,
    c) a plan.
       No hidden detail is required. Show only one M12 fixing bolt. Include the functional dimensions and indicate all surfaces that require machining.

70

11. Draw to a scale of twice full size in third-angle projection the following views of the bracket shown in fig. 10.15, including the fillets necessary for casting:
   a) a sectional front view on the main vertical centre line,
   b) a sectional plan on the main horizontal centre line,
   c) an end view.

   Select the datum surfaces and fully dimension the bracket. The size tolerance of the hole is to be 20 H7 and its curved surface is required to lie between two cylindrical surfaces coaxial with each other and a radial distance of 0.03 mm apart. The top surface, marked 'A', is required to lie between two parallel planes 0.08 mm apart, parallel to the base surface marked 'B'.

   Tracing paper may be used for a full-size drawing.

12. Figure 10.16 shows the oblique-projection assembly drawing of a clamping unit which is symmetrical about the cutting plane YY. This assembly consists of the following parts: (1) the lower casting with a central tangential web to support the 30 mm diameter boss; (2) the upper casting pivoted about the pin; (3) the 10 mm diameter pin, 45 mm long with a 20 mm diameter head, 5 mm long including 1 x 45° chamfer; (4) the washer with a circlip suitable for fastening the washer to the pin.

   Draw full size in first-angle projection, with all parts assembled and including the 35 mm long M10 bolt and a length of 30 mm diameter shaft,
   a) a sectional front view on vertical cutting plane YY,
   b) a sectional plan view on horizontal cutting plane XX,
   c) an end view.
   Include a parts list and a balloon reference system.

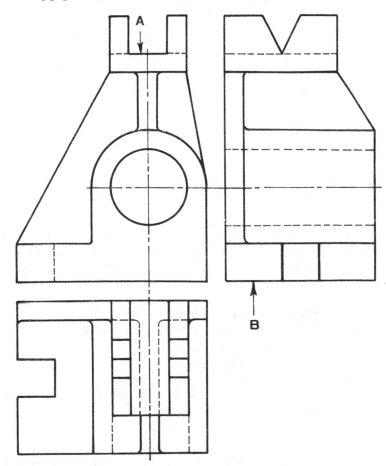

Fig. 10.15 Bracket

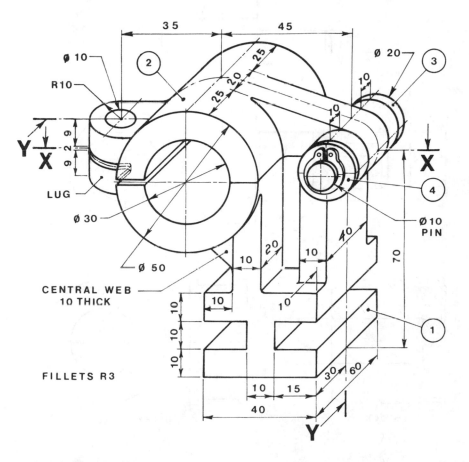

Fig. 10.16 Clamping unit

71

13. Draw a front view and an end view of the pivot pin shown in fig. 10.17. Scale dimensions where necessary and fully dimension the drawing. Longitudinal size tolerances are to be 0.3 mm. Fits for functional diameters are designated 12 n6 and 18 e9. The 12 mm and 18 mm diameter surfaces are to be concentric, i.e. the axis of the 12 mm diameter cylindrical portion is required to be contained within a cylinder 0.05 mm diameter co-axial with the 18 mm diameter portion.

As an additional exercise, fully dimension the pulley.

14. Figure 10.17 shows part of a slow-motion mechanism. Tolerance all indicated dimensions by showing the maximum and minimum size limits:

    fit between pulley and pivot pin to be 18 H8/f7,

    fit between pivot pin and block to be 12 H7/p6,

    fit between pulley and pivot-pin head to have a maximum axial clearance of 0.6 mm and a minimum clearance of 0.2 mm as shown.

Include the machining symbols and indicate functional dimensions by the letter F, one non-functional dimension by the letters NF, and an auxiliary dimension by the letters (AUX).

Tracing paper may be used.

15. Figure 10.18 shows the components comprising a small vice. The fixed jaw (1) is secured to the base plate (3) with two M6 countersunk-head screws (7). The moving jaw (2) is free to slide along the dovetail of the base plate (3). The spindle (4) with its handle (5) is located axially in the fixed jaw (1) by means of the collar (6) and two grub screws (8).

Place tracing paper over the views of the components and build up an assembly drawing by tracing each part in turn. Drawing instruments should be used, and the following views are required in third-angle projection with the faces marked 'B' touching:

a) a sectional front view in the direction of arrow A along the length of item (4), showing the handle (5) longitudinally;

b) a plan view;

c) an end view including hidden detail.

Balloon reference the assembly and present a suitable parts list, including the correct projection symbol.

16. a) Draw a front view and a sectional end view through the 6 mm diameter hole of item (4) in fig. 10.18. Scale dimensions where necessary and fully dimension the drawing to comply with the following conditions: the 16 mm diameter part to be cylindrical within 0.08 mm and the fits to be 12 n6, 16 f7, and 6 H9.

Include three machining symbols for the functional surfaces, which are in contact with sliding and rotating surfaces of the adjacent components when assembled.

b) As a further exercise, fully dimension the remaining components with all surfaces marked 'B' to be square with surfaces marked 'C' within 0.06 mm and the surfaces marked 'D' to be parallel to the surfaces marked 'C' within 0.05 mm.

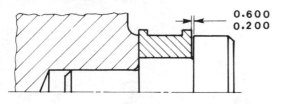

ASSEMBLY

0.600
0.200

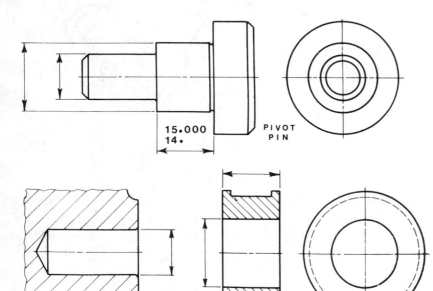

PIVOT PIN

15.000
14.

BLOCK

PULLEY

Fig. 10.17 Part of a slow-motion mechanism

72

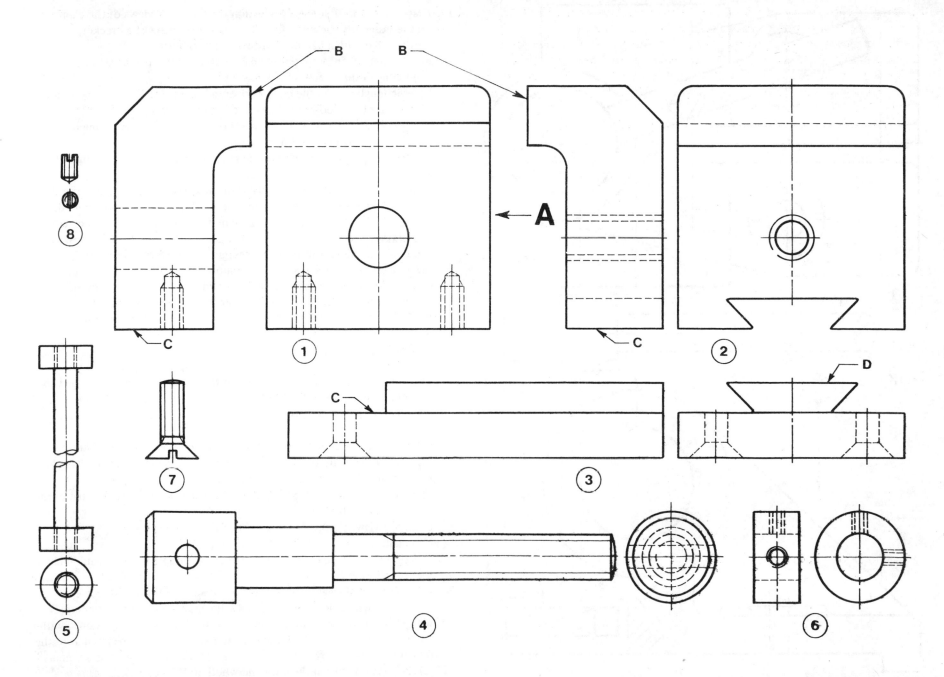

Fig. 10.18 Small vice

73

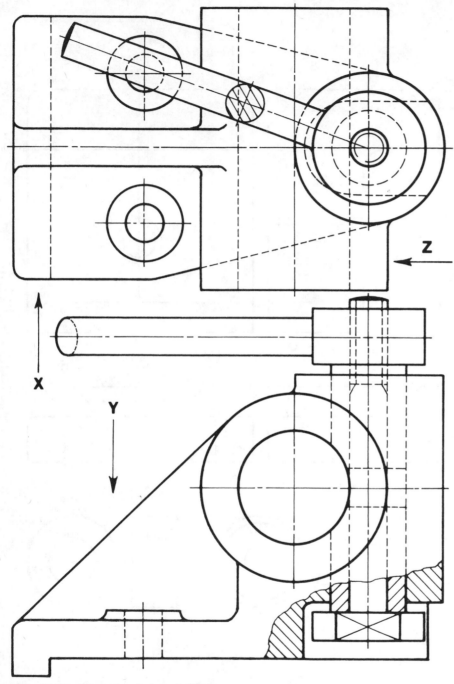

Fig. 10.19 Clamp bracket assembly

17. Draw full size in first-angle projection the following views of the clamp-bracket assembly shown in fig. 10.19, which consists of a bracket, a clamping bar with a handle, and two identical bushes:
    a) a sectional front view on the main vertical centre line of symmetry, looking in the direction of arrow X;
    b) a sectional plan on the main horizontal centre line through the 30 mm diameter hole, looking in the direction of arrow Y;
    c) a sectional end view on the vertical centre line of the clamping bar, looking in the direction of arrow Z.
    The clamping-bar handle may be positioned along the centre line of symmetry.
    Balloon reference the assembly with the parts list, and include the title block and the main functional dimensions. Tracing paper may be used.

18. Figure 10.19 shows an assembly drawing for a clamp bracket which consists of a bracket, a clamping bar with a handle, and two identical bushes.
    Prepare detailed working drawings, giving toleranced dimensions, material requirements, machining symbols, and instructions for manufacture.
    Provision should be shown for preventing the bushes from rotating.
    The bracket hole of 30 mm diameter is to have H7 fit, and the fits are to be H7/k6 between the bracket and bushes and H8/f7 between the spindle and the bushes.
    Tracing paper may be used.

19. Figure 10.20 shows the components comprising a low-speed pulley unit. The bush is a press fit within the wheel, and the assembly is to be arranged so that both faces marked 'X' and both faces marked 'Y' are adjacent respectively. The spindle is held in the bracket with a hexagon nut and washer.
    Place tracing paper over the views of the components and build up an assembly drawing by tracing each part in turn. Drawing instruments should be used, and the following views are required:
    a) a sectional front view with the cutting plane along the centre line of the spindle assembly;
    b) an end view projected to the right of the front view;
    c) a plan.
    Balloon reference the assembly, present a suitable parts list, and include the correct projection symbol.

20. Draw a sectional view through the bush, item 3, in fig. 10.20. Scale dimensions where necessary, and fully dimension the drawing to comply with the following conditions: the axis of the hole is to be contained within a tolerance cylinder of 0.02 mm diameter and is to be perpendicular to the datum face marked 'X'. The fit between the spindle and bush is designated 20 H9/e9 and between the bush and wheel 30 H7/p6.
    Add the appropriate machining symbol(s).

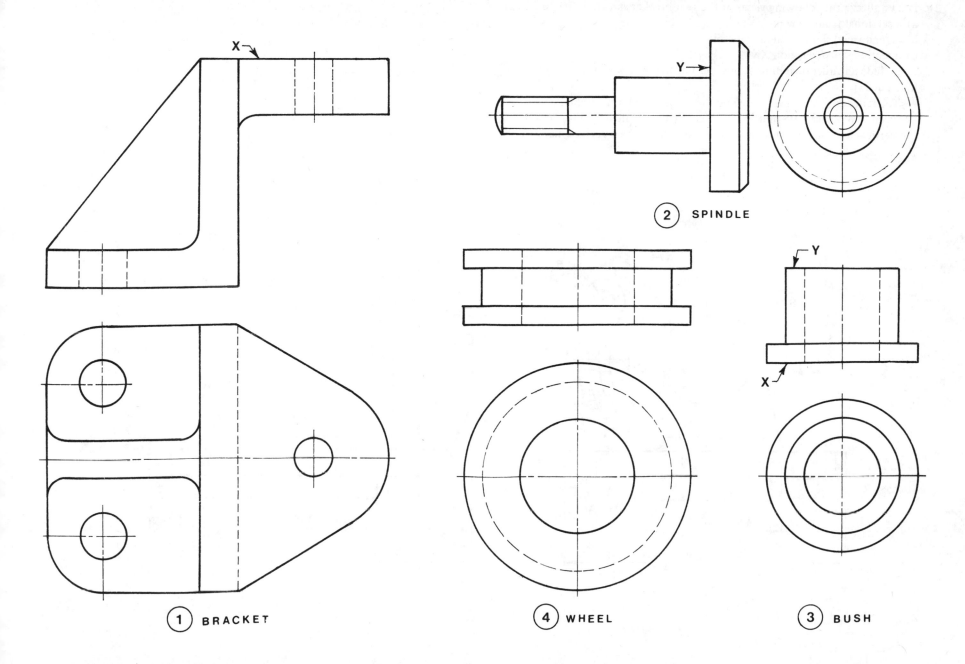

① BRACKET

② SPINDLE

④ WHEEL

③ BUSH

Fig. 10.20 Low-speed pulley unit

21. Draw full size the following views of the gear bracket shown in fig. 10.21:
    a) a sectional front view on YY;
    b) an end view, showing hidden detail;
    c) a sectional plan view on XX.

    Include all functional dimensions and indicate which surfaces should be machined.

    Tolerance for squareness, showing the axis of the horizontal hole to lie between two planes 0.05 mm apart which are perpendicular to the axis of the vertical hole and at the same time parallel to the base surface marked 'A'.

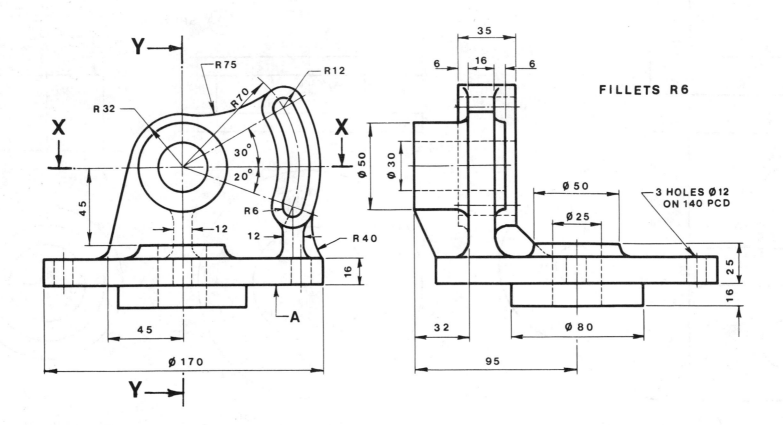

Fig. 10.21 Gear bracket

22. Trace or redraw separately each item shown in fig. 10.22. Consider the four washers (6), (7), (8), shown with the taper pin, and (10) to be of the same standard dimensions. Include a locking arrangement for the nut (9).

  Dimension each part completely with the following fits of nominal size of 8 mm diameter: (a) H7/g6 fit between fork (2) (consisting of two plates and a 20 mm × 20 mm square boss) and spindle (4) (65 mm long with 8 mm end diameters and middle part of 20 mm diameter); (b) H9/e9 fit between wheel (5) and axle (3); (c) H7/k6 fit between spindle (4) and supporting attachment (11).

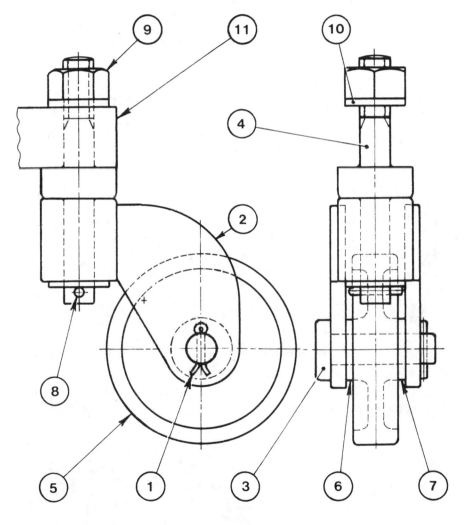

Fig. 10.22 Heavy-duty castor

23. The designer's detail layout sketch of a jig used for inspecting shafts is shown in fig. 10.23, ready for a general-assembly drawing to be drawn.

  Draw full size the following views of the assembled jig:

a) a sectional front view on the vertical cutting plane AA,

b) an end view,

c) a plan with all hidden detail to be shown.

  The jig is to be assembled with a 150 mm length of 20 mm diameter shaft clamped between the jaw and the vee groove in the base. The shaft is to be shown protruding 10 mm beyond the right of the base.

  Include a title block, a parts list, and a balloon reference system and suggest suitable materials.

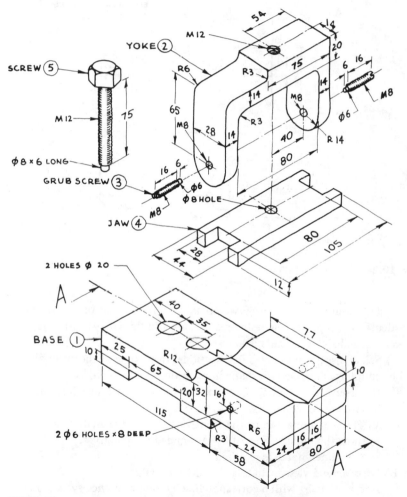

Fig. 10.23 Jig

24. The flange marked 'A' in fig. 10.24 is to accommodate a 10 mm x 10 mm square key and a 40 mm diameter 150 mm long shaft.

Select a scale and draw (a) the flange, (b) the square key, and (c) the shaft, and then tolerance them according to the following fits: H7/h6 fit between the flange and the shaft; H7/k6 fit between the key and the flange and between the key and the shaft.

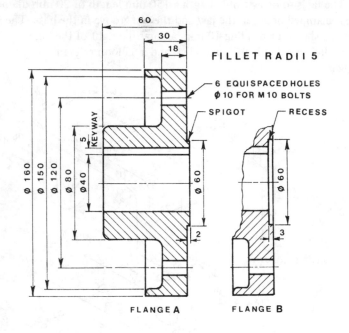

FILLET RADII 5

6 EQUISPACED HOLES
Ø 10 FOR M 10 BOLTS

SPIGOT   RECESS

FLANGE A   FLANGE B

Fig. 10.24 Flange coupling

25. A shaft coupling connecting two co-axial shafts consists of two flanges, identical except for spigot and recess shown in fig. 10.24; two keys; two shafts; and six bolts, nuts, and washers.

a) Select a scale and draw a half-sectional front view of the assembled coupling, showing two shafts in position. The required view should show the assembled coupling in section above the shaft centre line and it must include one of the bolts.

The remaining part of the front view, below the shaft centre line, is to show the outside view of the assembled coupling.

No hidden detail is required.

b) Draw an end view including all hidden detail.

Add a parts list with a corresponding balloon-reference system and suggest suitable materials.

26. Draw the following views of the bearing-and-bracket assembly shown in fig. 10.25:

a) a sectional front view, viewed in the direction of arrow B, through the centre of the vertical web;

b) a half-sectional end view on AA, the half-section to be to the right of the vertical centre line of symmetry.

Include one M12 nut, washer, and bolt.

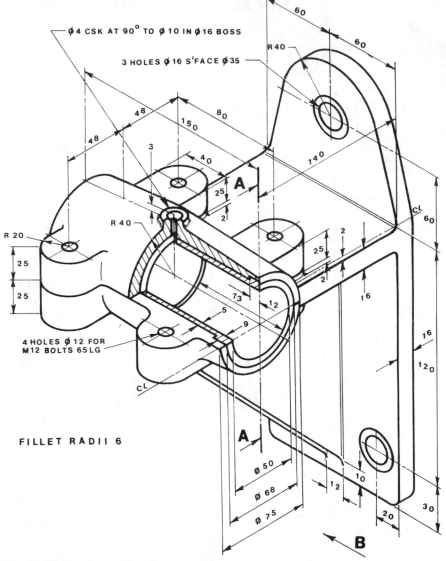

FILLET RADII 6

4 HOLES Ø 12 FOR
M12 BOLTS 65 LG

Fig. 10.25 Bearing-and-bracket assembly

27. Figure 10.26 shows five detail parts which together make up a toolmaker's vice. This consists of a body (1), a moving jaw (2), a bottom plate (3), a screw (4), a special grub screw (5), and two M4 countersunk screws, 16 mm long, which are not shown.

Select a scale and draw in correct projection the following views of the fully assembled toolmaker's vice:

a) a sectional front view, taking the screw axis to be a cutting plane;

b) an end view;

c) a plan.

Show all hidden detail in views (b) and (c).

Include a title block with all relevant information and add a parts list with a balloon-reference system.

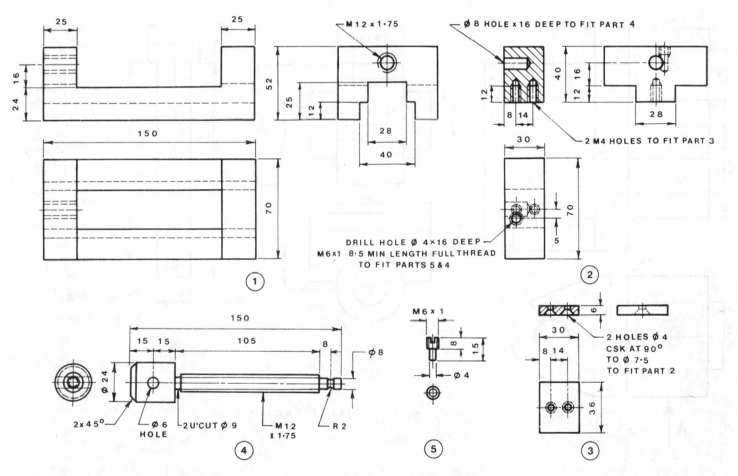

Fig. 10.26 Toolmaker's vice

28. Figure 10.27 shows a Hooke's coupling, which is used to connect two shafts that have a small degree of misalignment. It consists of two forks (1), one centre (2), two pins (3), and two collars (4).

Select a scale and draw the following views with all parts correctly assembled:

a) a sectional front view in the direction of arrow A,
b) a plan view,
c) an end view showing hidden detail.

Each collar is to be secured by a tapered 3 mm diameter pin, and each fork is to be secured to its shaft by a 10 mm x 10 mm key.

Include a title block, add a parts list with the reference balloons, and suggest suitable materials to be used.

29. Figure 10.27 shows a Hooke's coupling. The centre (2) basically consists of two tubular pieces cast together at 90° to each other.

Select a scale and draw the centre and fully dimension it, satisfying the following conditions:

a) the fit between the centre and the pins (3) is to be H7/n6,
b) the axis of the vertical hole is required to lie between two parallel planes 0.04 mm apart which are perpendicular to the axis of the horizontal hole.

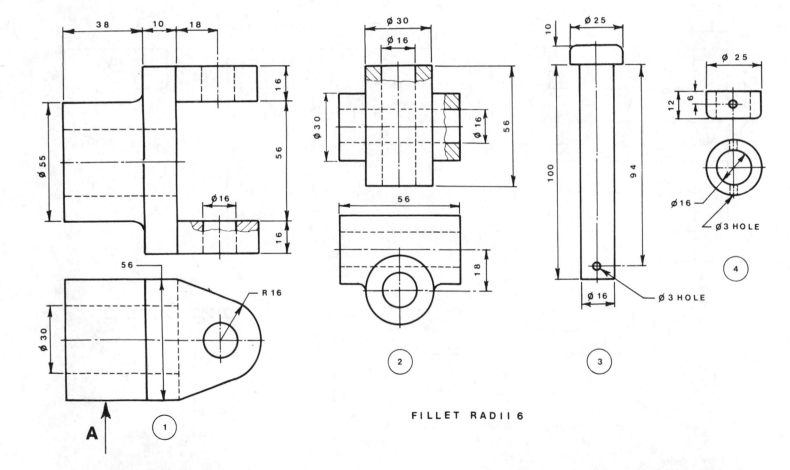

FILLET RADII 6

Fig. 10.27 Hooke's coupling

80

30. The adjusting unit shown in fig. 10.28 consists of a body (1), a handle (2), a pivot (3), a screw (4), a collar (5), and a pin (6).

Draw the following views of the completely assembled adjusting unit:

a) a front view in the direction of arrow B, including all hidden detail;

b) a sectional end view on the cutting plane AA.

Include the main functional dimensions that have a bearing on the assembly. Insert a title block, with the relevant information, and a parts list with the corresponding reference balloons and suggest suitable materials.

31. Figure 10.28 shows the adjusting unit of an instrument used for electrical adjusting purposes. By moving the handle (2), the required position of the slot in the pivot (3) is obtained.

Draw twice full size the pivot, incorporating the following conditions:

a) a fit of 8 H7/h6 is required between the pivot and the body (1), and a fit of 10 H7/p6 between the pivot and the handle (2);

b) the axis of the pivot is required to lie between two parallel planes 0.06 mm apart which are symmetrically disposed about the median plane of the slot in the end of the pivot.

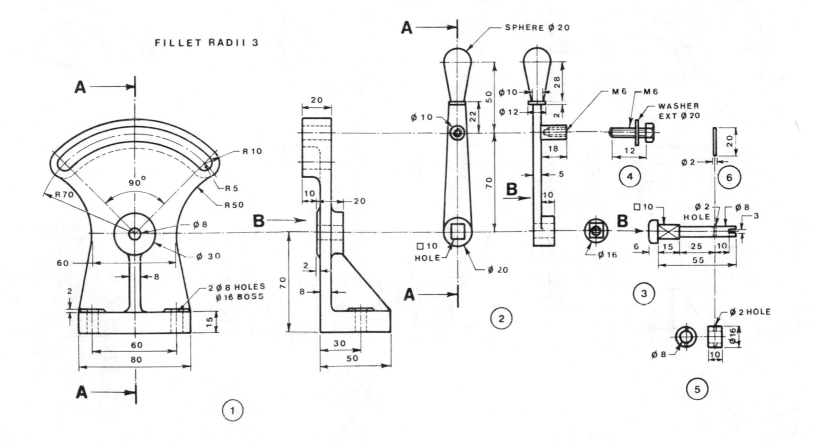

Fig. 10.28 Electricity adjusting unit

81

32. Figure 10.29 shows the detail parts of a right-angle cock, in which water passes through the hole in the plug, fitted in the body. Rotation of the plug through a right-angle stops or allows the flow of water. The right-angle cock consists of one body (1), one plug (2), one gland (3), one packing (4), and two M12 studs (5), 40 mm long and 13 mm full-thread length at each end, with two nuts 10 mm thick and two washers 2.5 mm thick (not shown).

Select a scale and draw the following views in first-angle projection with all parts assembled and so arranged that there is a free passage for flowing water:

a) a sectional front view on YY,
b) an end view including hidden detail,
c) a plan.

Insert a title block with reference balloons and a parts list, with suggested materials.

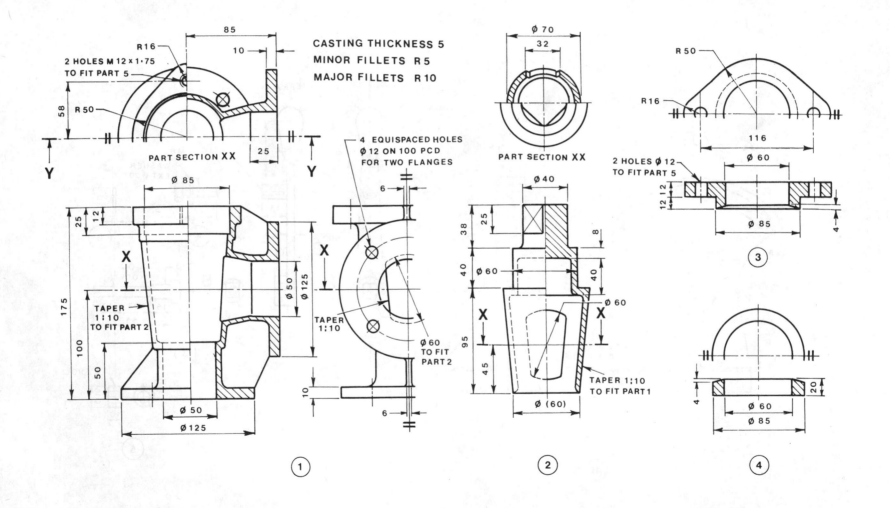

Fig. 10.29 Right-angle cock

82

33. The component parts of a tailstock are shown in fig. 10.30. They consist of a body (1); a barrel (2); a cap (3); a spindle (4); a handwheel (5); a centre (6); a key (7); an M10 hexagon-head screw, 50 mm long with 20 mm minimum length of full thread to fit part (1); and an M10 nut, 8 mm high with a 2 mm thick washer to fit part (4).

With all parts correctly assembled, draw full size the following views:

a) a sectional front view on AA,

b) an end view,

c) a plan.

Include a title block with all relevant information and reference balloons with a parts list, suggesting suitable materials.

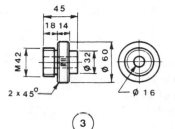

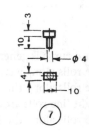

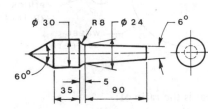

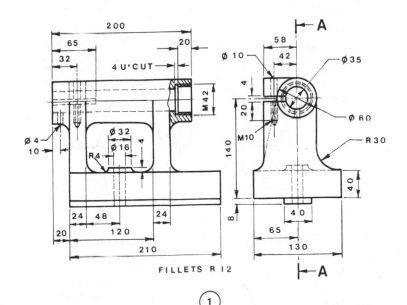

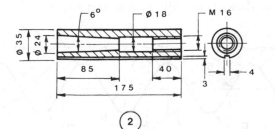

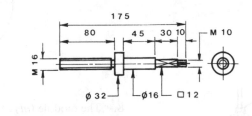

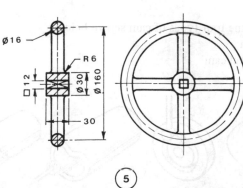

Fig. 10.30 Tailstock

83

Self-assessment questions

1. A rod EF is free to move in such a way that E is always in contact with a line OY and F is always in contact with another line OX.
   Select from the diagrams shown below the locus of the mid-point P.

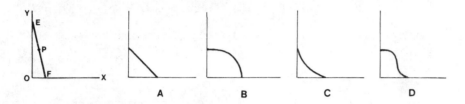

2. An involute curve is generated when:
   A a circle rolls along a straight line without slip.
   B a circle rolls along the outside of an arc.
   C a straight line rolls around a circle without slip.
   D a circle rolls along the inside of an arc.

3. Match the thread features shown below with their correct names:
   a) crest
   b) flank
   c) thread angle
   d) pitch

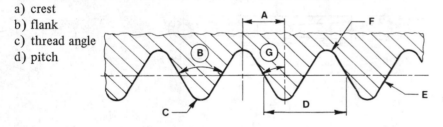

4. Select from the four types of cam shown below:
   a) a disc cam
   b) a cylindrical cam

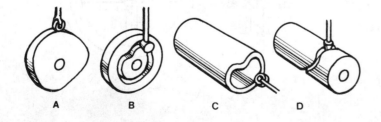

5. Four diagrams shown below indicate the different motions for a cam-follower rise. Which one shows a uniform acceleration?

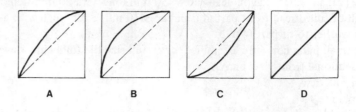

6. Match the involute gear features shown below with their correct names:
   a) tooth thickness
   b) dedendum
   c) tip

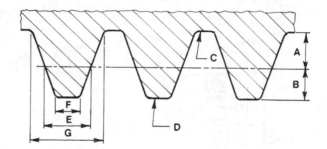

7. For the three gears shown in mesh, the gear ratio is:
   A 8:1
   B 4:1
   C 2:1
   D 1:2

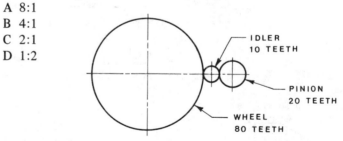

8. The module (m) of a gear is the ratio of:
   A the number of teeth to the pitch-circle diameter.
   B the addendum to the dedendum.
   C the dedendum to the addendum.
   D the pitch-circle diameter to the number of teeth.

9. Four types of rolling bearing are shown below. Which bearing will support only journal loads?

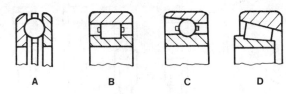

10. Which of the bearings shown below is suitable for supporting the axial thrust load shown?

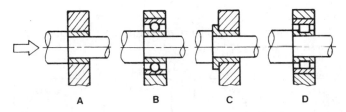

11. Using BS 4500A, select the correct maximum and minimum size limits to give the Ø18 H8/f7 fit.

a) Maximum limits
b) Minimum limits

A 18.033/17.980
B 18.027/17.984
C 18.000/17.966
D 18.000/17.959

12. From the list shown below, select the geometrical-tolerance symbols belonging to the following types of tolerance:
a) attitude
b) form
c) location

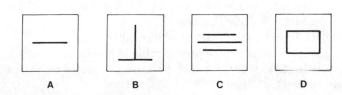

13. Match the tolerance symbols shown below with their correct names:
a) roundness
b) cylindricity
c) concentricity

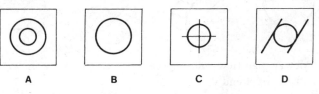

14. Which of the properties listed below is most appropriate when considering wrought iron for making chains for lifting purposes?
A Hardness
B Brittleness
C Castability
D Toughness

15. Of the assemblies shown below, which uses the correct conventional representation according to BS 308:1972?

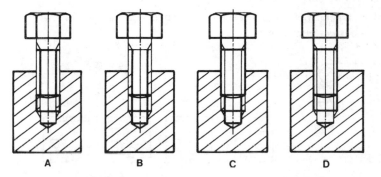

16. The two components shown in section are joined together by means of:
A a bolt
B a stud
C a screw.

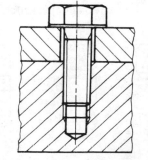

85

17. Which of the keys shown below is most suitable for a keyway in a tapered part of a shaft?

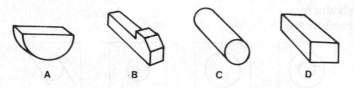

18. A front view, a plan view, and a space for an end view are shown below. Which is the correct end view?

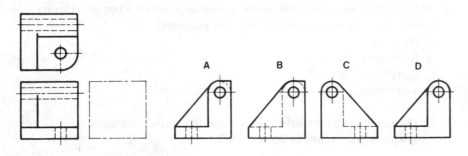

19. A front view, an end view, and a space for a plan view are shown below. Which is the correct plan view?

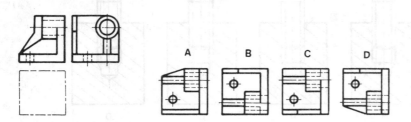

20. The following conventional representations are shown below: internal screw thread, diamond knurling, interrupted view, spur gear, and bearing. Which uses the correct conventional representation according to BS 308:1972?

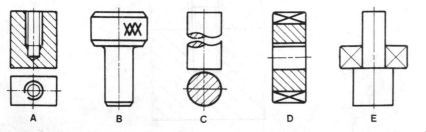

21. Which is the correct sectional view on XX below?

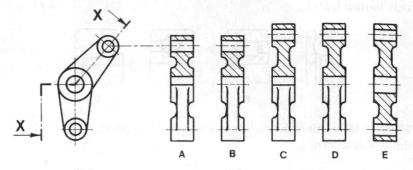

22. Which is the correct conventional representation of the splined hole according to BS 308:1972 below?

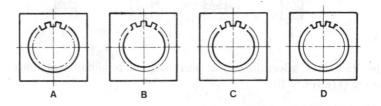

23. Each drawing of a component shown below is in orthographic projection and consists of a front view, an auxiliary view, and a projection symbol. Which drawing is correct?

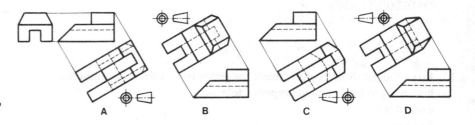

24. Which of the following statements is true?
   A Balloon reference numbers must always be positioned around a component in an increasing or decreasing order.
   B Assembly drawings should never include any dimensions.
   C For small sub-assemblies the parts list should be excluded.
   D For very large assemblies the parts list is usually placed on a sheet separate from the drawing.

86

Selected solutions

Solutions to some of the questions can readily be found from the text. Solutions to a selection of other questions are given here, though in some cases only partial solutions are shown.

All diagrams are drawn to a reduced scale, and, for simplicity, the fillets and some detail features on certain drawings have been omitted.

All stud, bolt, and nut assemblies are drawn using the approximate method explained in *Engineering drawing for technicians volume 1*.

Page 7, Q 13

Page 7, Q 14

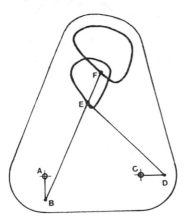

Page 7, Q 15

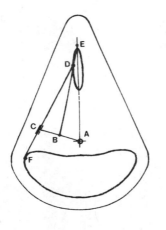

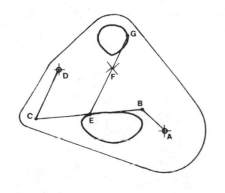

Page 22, Q 9

Page 22, Q 10

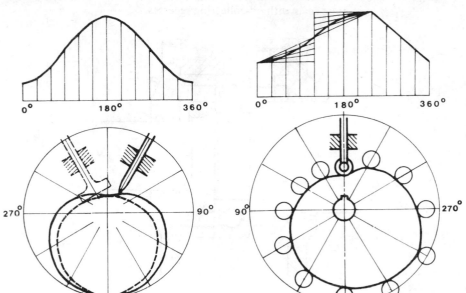

Page 32, Q 16   320 mm

Page 32, Q 18   (a) 648 mm, 216 mm; (b) 672 mm, 240 mm; (c) 618 mm, 186 mm; (d) 37.7 mm; (e) 18.85 mm; (f) 432 mm

Page 32, Q 19   (a) 96, 24; (b) 672 mm, 168 mm; (c) 420 mm; (d) 686 mm, 182 mm; (e) 654.5 mm, 150.5 mm; (e) 3.75°, 15°

Page 32, Q 20   (a) 16, 12; (b) 256 mm, 192 mm; (c) 288 mm, 224 mm; (d) 216 mm, 152 mm; (e) 240.56 mm, 180.42 mm; (f) 32 mm; (g) 224 mm

Page 53, Q4

a)	hole	50.160	shaft	49.870
		50.000		49.710
b)	hole	100.035	shaft	100.045
		100.000		100.023
c)	hole	150.040	shaft	150.125
		150.000		150.100

Page 53, Q6   9.96
9.92

Page 41, Q 19 Many correct solutions are possible. This solution shows gap-type seals with oil-collecting recesses.

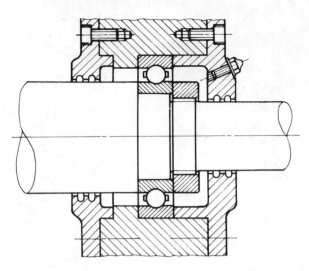

Page 67, Q 1

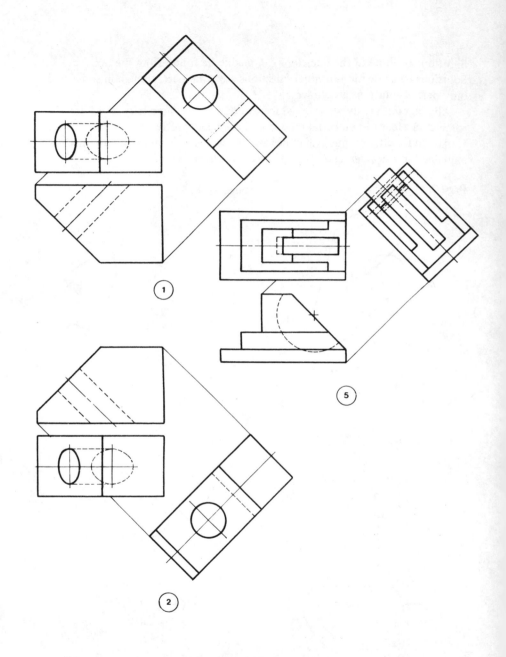

Page 54, Q 18

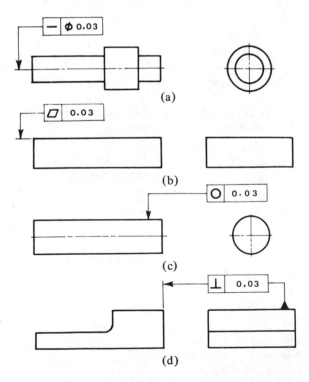

(a)

(b)

(c)

(d)

88

Page 68, Q 2

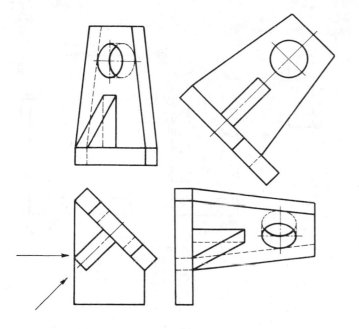

Page 68, Q 3

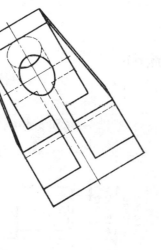

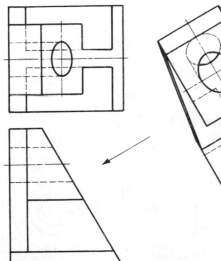

Page 69, Q 4

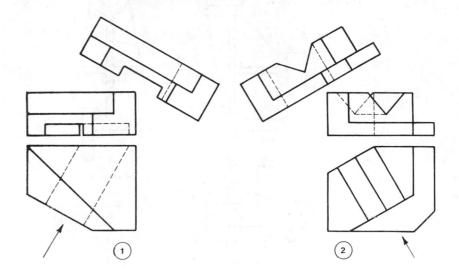

Page 69, Q 5

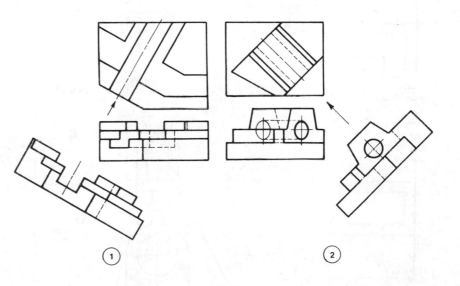

Page 70, Q 10

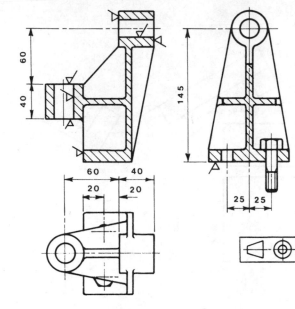

Page 71, Q 12

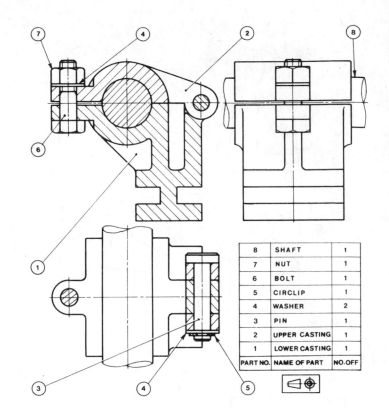

PART NO.	NAME OF PART	NO. OFF
8	SHAFT	1
7	NUT	1
6	BOLT	1
5	CIRCLIP	1
4	WASHER	2
3	PIN	1
2	UPPER CASTING	1
1	LOWER CASTING	1

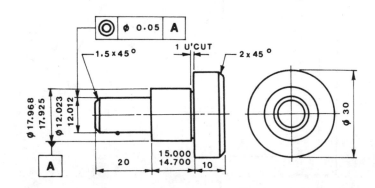

Page 71, Q 11

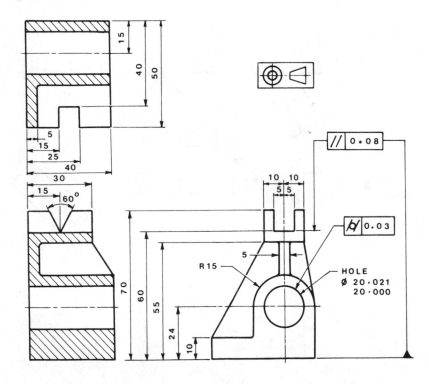

Page 72, Q 13

90

Page 72, Q 14

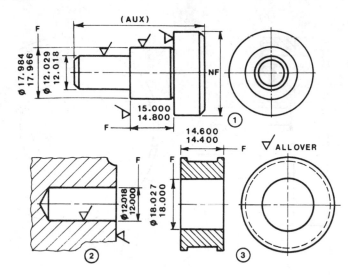

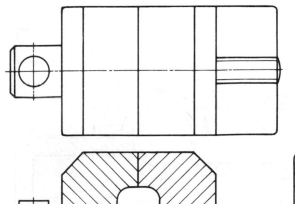

Page 72, Q 15

Page 72, Q 16

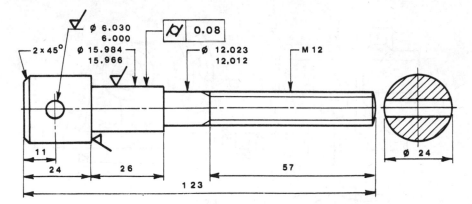

Page 74, Q 19

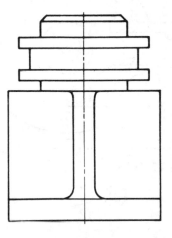

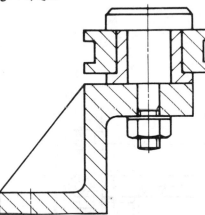

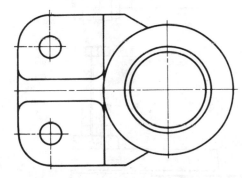

NO.	ITEM	MATERIAL	NO, OFF
1	BRACKET	CAST IRON	1
2	SPINDLE	STEEL	1
3	BUSH	BRONZE	1
4	WHEEL	DURALUMIN	1
5	NUT	STEEL	1
6	WASHER	STEEL	1

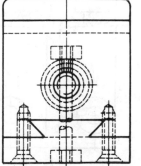

Page 74, Q 20 See page 63, fig. 10.1.

Page 76, Q 21

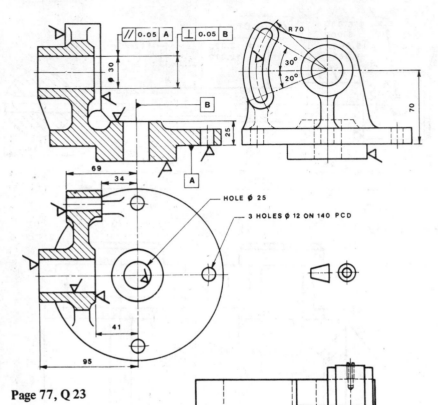

HOLE Ø 25

3 HOLES Ø 12 ON 140 PCD

Page 77, Q 23

Page 78, Q 25

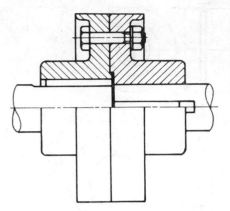

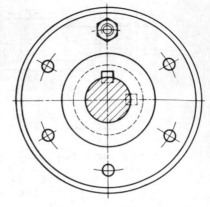

Page 78, Q 26

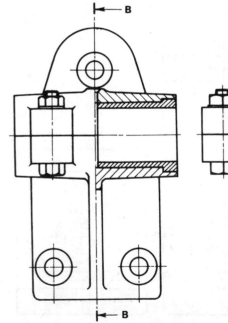

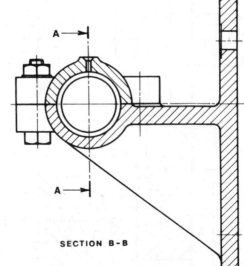

SECTION B-B

Page 79, Q 27

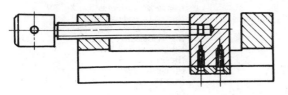
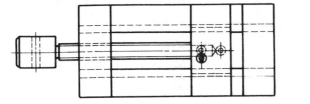
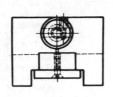

Page 80, Q 28

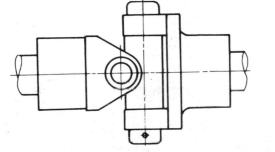
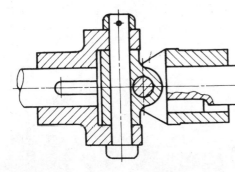
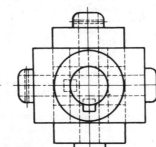

Page 81, Q 30

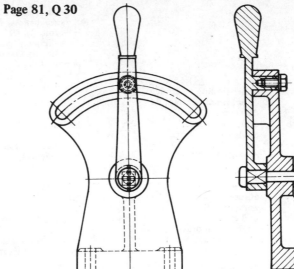
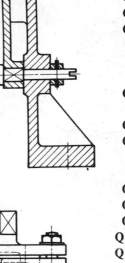

Page 82, Q 32

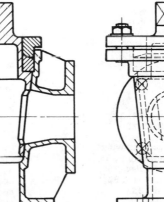
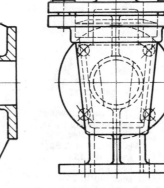
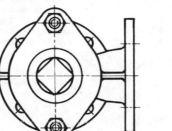

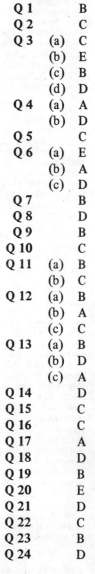

Index